Induction generators for wind power

The book cover: GE wind turbines in Anrenburg, Germany. Photo copyright GE Wind Energy. Reproduced with permission from GE Wind Energy.

Information contained in this work has been obtained by Multi-Science Publishing Company, Ltd., from sources believed and reliable. However, neither Multi-Science Publishing Company nor its authors nor its editors guarantee the accuracy or completeness of any information published herein and neither Multi-Science Publishing Company nor its authors shall be responsible for any errors, omissions, or damages arising out of use of this information. This work is published with the understanding that Multi-Science Publishing Company and its authors are supplying information but are not attempting to render engineering or other professional services. If such services are required, the assistance of an appropriate professional should be sought.

Dedicated to C. E. Andersen

Wind turbine nacelles ready for delivery. Photo copyright Vestas Wind Systems. Reproduced with permission from Vestas Wind Systems.

Table of Contents

Lifting up the rotor. Photo copyright GE Wind Energy. Reproduced with permission from GE Wind Energy.

Preface

This book is concerned with understanding, modelling, grid-connection and fault ride-through capability, i.e. uninterrupted operation without sub-sequent disconnection at grid disturbances, of wind turbines equipped with induction generators which may be fixed-speed or variable-speed and converter controlled. This issue is very timely due to the rapid incorporation of wind power into power systems across the globe, and relevant for maintaining the reliable operation of power systems.

Electrical power supply technology is well-known and based on the generation and controllability of conventional power plants and their synchronous generators. Such conventional power plants provide power-frequency and voltage control. This well-known centralised power supply technology is giving way to partly-unknown technology of electrical power supply that is wind power. Intensive incorporation of wind power into the power grids is a relatively new issue. Starting in the late 1990s in Denmark, an ambitious governmental programme for commissioning of large offshore windfarms and subsiding local wind power and combined heat-power units was followed by an exponential growth of wind power in for example Denmark, many European countries, USA, Canada, Australia, India and Brazil.

The commissioning of large (offshore) windfarms containing hundreds of wind turbines is seen in many countries. The rating of large windfarms require that the windfarms feed directly the transmission power grids. In selected offshore sites in the North Sea, there are plans to establish about 30 GW of wind power capacity to be connected to the transmission power grids of Denmark, Germany and The Netherlands, interconnected by an offshore transmission grid. Such incorporation of wind power into transmission power grids may require that conventional power plants reduce their part in the power generation mix in favour of wind power - but this then introduces a challenge with regard to maintaining power system stability. Such power system stability topics are various and include active power balance and control due to the fluctuating nature of wind, the identification of bottlenecks in the transmission grid for transporting power from windfarms to consumption centres, power quality issues due to emission flicker caused by wind turbines, short-term voltage stability and fault ride-through capability of wind turbines to minimise the requirements for immediate power reserves. Therefore, it focuses on development of dynamic models of different wind turbine concepts and the understanding of the interaction between wind turbines and transmission power grids.

A common question has been: "What happens when the power grid with grid-connected wind turbines is subject to a short-circuit fault?" This book answers this question and brings understanding with regard to dynamic models of wind turbines applied in investigations of short-term voltage stability. As about 85% of the wind turbine market comprises wind turbines equipped with either fixed-speed or with variable-speed, converter-controlled induction generators, this book concentrates on induction generator-based wind turbines. As the power rating of wind turbines rapidly increases and power electronics converters become commonplace in generator control, this book presents details about power electronics converters' modelling with regard to short-term voltage stability, control, protection and fault ride-through capability.

The book is primarily addressed to students and post-graduate students with an interest in the electrical power supply and grid-connection of wind turbines, electrical power engineers and consultants working on the grid-connection of windfarms and modelling of wind turbines, electrical power companies and transmission system operators considering commissioning of a large wind-farm, wind turbine manufacturers when concerning with wind turbine modelling and the documentation of fault ride-through operation of the wind turbines, and any other with an interest in the technical issues within wind power.

The author is thankful to the American wind turbine manufacturer GE Wind Energy, the Danish wind turbine manufacturer Vestas Wind Systems and the manufacturer Siemens (former Bonus Energy) for offering photographs and providing technical information. The author is also thankful to Red Eléctrica de España, transmission system operator of Spain, for information about commissioning and operation of wind power in the Spanish power system, Dr. Sigrid M. Bolik, with Vestas Wind Systems, for sharing the results of her Ph.D. project, Prof. Trevor J. Price, with School of Technology, University of Glamorgan, for editorial review, and Prof. John Twidell, the Editor-in-Chief of Wind Engineering, for encouraging to write this book.

The author is also thankful to his former employer Energinet.dk, Transmission System Operator of Denmark for natural gas and electricity, for their support.

Vladislav Akhmatov
Fredericia, Denmark

1 Introduction

Modern power systems in the industrialised world are characterised by massive incorporation of electricity-producing wind turbines. This development is the result of successfully developing wind technology, governmental targets, and subsidies for, the incorporation of more renewable energy and a stronger demand for reducing pollution and improving the environment.

Electricity-producing wind turbines are today the largest source of environment-friendly energy production as they have gained the largest power efficiency compared to other forms using natural free-of-charge energy sources such as tidal, waves and sun. The maximum power efficiency of electricity-producing wind turbines can be in the range of 50% and still competitive to the conventional power generation (Hansen et al., 2004(a)).

Modern industrialised power systems are based upon centralised, conventional power plants that control the grid voltage and provide balance between power generation and consumption. Reliable and safe operation of these power systems is based on the operation and control of conventional power plants, which is a well-known and developed technology (Kundur, 1994).

Conventional power plants are traditionally synchronous generator based. When the power grid is subject to a short-circuit fault, the excitation control of conventional power plants contributes to the voltage re-establishment in the power grid and their frequency control ensures grid frequency during such events.

Electricity-producing wind turbines are an alternative and promising, but still partly unknown, technology with regard to their impact on the operation and stability of the power grids into which they feed. Furthermore, the majority of wind turbines are equipped with induction generators of several different concepts. The progressive incorporation of wind power into power grids reduces the power supply from, and the share of, the conventional power plants in favour of wind power. This process corresponds to a transition from the well-known and developed technology of the power grid operation with the use of conventional power plants, towards the partly-unknown technology of wind power. This creates several issues regarding maintaining power system stability and the representation of electricity-producing wind turbines in stability investigations (Santes, 2002; McArdle, 2004).

The main goal of this book is to provide an introduction to the modelling of electricity-producing wind turbines equipped with induction generators of several different concepts, their control and the fault ride-through solutions.

1.1 Book outline

This introductory part gives an overview of wind power growth in power grids world-wide. As the amount of the wind power rapidly increases and the power generation in conventional power plants is reduced in favour of wind power, the modelling of electricity-producing wind turbines becomes a very relevant issue for investigations of power system stability.

The majority of electricity-produced wind turbines are today equipped with induction generators of different concepts. **Chapter 2** presents wind turbine concepts based on such induction generators.

Modern wind turbines are complex electromechanical systems. Electromechanical interaction between wind turbines and power grids may be excited by grid faults (Akhmatov et al., 2000(a)). **Chapter 3** addresses the wind turbine mechanical construction with regard to investigations of power system stability. **Chapter 3** covers the basics of the rotor aerodynamics, the shaft representation and the blade-angle control for wind turbine modelling in power system stability.

Chapter 4 explains the modelling details of conventional induction generators with a short-circuited rotor circuit. Such conventional induction generators are used in fixed-speed wind turbines. This chapter is coupled with **Chapter 5** describing the issues of short-term voltage stability and of fault-ride-through capability of fixed-speed wind turbines. **Chapter 5** also presents an example of the detailed modelling of a large (offshore) windfarm consisting of eighty 2 MW wind turbines.

Chapter 6 presents the concept, the modelling details and the ride-through capability of pitch-controlled wind turbines equipped with induction generators with dynamic rotor resistance. This concept relates to the Vestas OptiSlip®[1] wind turbines.

Chapter 7 focuses on variable-speed, pitch-controlled wind turbines equipped with doubly-fed induction generators and partial-load frequency converters. At present, this concept becomes very relevant for large offshore projects world-wide, which is why the accurate modelling of, and fault-ride-through solutions for, this wind turbine concept are in focus. **Chapter 7** gives a general description of this concept and introduces solutions of the leading wind turbine manufacturers on the fault-ride-through capability and on the ancillary (system) control.

The concept of induction generators with full-rating frequency converters is shortly described in **Chapter 8**. This concept is interesting because it gives an example of direct-current (DC-) connection of an induction generator based wind turbine to the AC- power grid.

Chapter 9 discusses the model aggregation of large windfarms containing a large number of wind turbines into an equivalent with a single machine or with fewer machines of rescaled power capacities. This issue is relevant for stability investigations carried out for large power grids that already have a significant number of component models such as power plants and control equipment, lines, transformers, and voltage- and frequency-dependent loads.

A new retro-fitting solution, e.g. a device incorporated into an older wind turbine originally without fault ride-through and providing the ride-through capability, called a Transient Booster[TM2] is presented in **Chapter 10**. This retro-fitting solution was announced in the year 2005.

A summary is given in **Chapter 11** presenting the lessons from dynamic wind turbine modelling for investigations of short-term voltage stability.

1.2 Wind power outlook

Commercial production of electricity-producing wind turbines started in Denmark in the early 1970s. At that time, production facilities were relatively small and the rated power of wind turbines was in the range of 20 kW. At around the same time, the incorporation of wind power into the Danish power grid began, **Figure 1.1**.

---

[1] OptiSlip is a registered trade mark of Vestas Wind Systems.
[2] Transient Booster is a trade mark of ABB Ltd., Zürich, Switzerland.

The incorporation of wind power in Denmark continued through the 1980s, whereas the rated power of wind turbines increased to hundreds of kW. In the early 1990s, the first mega-watt class wind turbine became available. Through the 1990s, Denmark (especially Western Denmark due to excellent wind conditions) experienced a drastic increase of wind power incorporation into the power system. At this time, the wind power share in the power generation mix has passed 10% which conventionally has been considered as the maximum limit of how much wind power a classical power system can bear. In 2005, the wind power share in the Danish power generation mix was about 20% and specifically in Western Denmark more than 30% (Eriksen et al., 2006(a)).

In 1997, the Danish government announced an ambitious programme to reach 4,000 MW wind power capacity in Denmark by the year 2030. This figure must be seen in relation to the minimum and maximum peak-loads of power consumption, which in the year 2005 were 1,200 MW and 6,300 MW, respectively (Eriksen et al., 2006(a)).

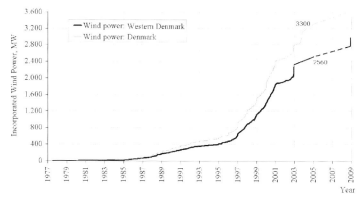

Figure 1.1 Incorporation of wind power in Denmark from 1977 with prognosis to 2009. Courtesy of Energinet.dk, Transmission System Operator of Denmark.

The Danish goal of wind power incorporation is to be reached by (i) commissioning large off-shore windfarms and (ii) the replacement programme (Eriksen et al., 2006(b)). The replacement programme means a growth in grid-connected wind power on-land due to the replacement of older, small wind turbines by new, larger ones. The replacement programme encourages 350 MW more wind power in Denmark.

Table 1.1 presents large offshore windfarms commissioned and announced in Denmark. The large offshore windfarms commissioned in Denmark are all with wind turbines rated at 2 MW and above. By the year 2010, there will be 430 MW more offshore wind power in Denmark.

The first Danish wind turbines were manufactured as fixed-speed units equipped with conventional induction generators, e.g. with short-circuited rotor circuits. Historically, the largest part of the wind turbines incorporated in Denmark have been such fixed-speed units.

During the 1990s, wind turbine manufacturing developed into professional businesses with the establishment of wind turbine manufacturing companies in Denmark as well as in the other European countries, the USA, and the development of new wind turbine concepts. Compared to the facilities of the early 1970s, the modern wind turbine manufacturing is the high-tech., competitive branch combining electric machinery, advanced mechanics and aerodynamics, power electronics and advanced control. **Table 1.2** lists the world's largest wind turbine manufacturers and their wind turbine concepts. Competition for large projects is tough, influencing the wind power market and the winning wind turbine concepts.

Modern commercially available wind turbines have rated powers in the range of 1.5 MW to 3.6 MW. Prototypes with rated powers approaching 5 MW were operated in the early 2000s. As expected, the first 10 MW wind turbine will be offered by the year 2010. **Figure 1.2** shows snap-shots from the wind turbine manufacturing processes kindly provided by GE Wind Energy. **Figure 1.3** shows assembling of the 3.6 MW GE wind turbines at Arklow, Ireland.

During the 1990s and early 2000s, many European countries, USA and Australia experienced massive incorporation of wind power into their power grids. In 2005, wind turbines supplied 25% of the electricity consumption in several German states and Spanish provinces. The largest incorporation of wind power in Europe will take place offshore in the North and the Baltic Seas (Source: European Wind Energy Association). **Table 1.3** provides some figures for these ambitious projects.

Offshore Windfarm	Location	Year of construction	Rated power (MW)
Middelgrund	Shore of Copenhagen	2001	40
Horns Rev A	North Sea	2002	160
Rodsand 1 / Nysted	South to Lolland	2003	165
Horns Rev B	North Sea	Expected by 2009	215
Rodsand 2	South to Lolland	Expected by 2010	215

Table 1.1   Large offshore windfarms in Denmark. Source: Danish Energy Authority (Energistyrelsen).

Manufacturer	Wind turbine concepts
Vestas Wind Systems	Fixed-speed with conventional induction generators.
	OptiSlip® with induction generators and adjustable rotor resistance.
	Variable-speed with doubly-fed induction generators.
GE Wind Energy	Variable-speed with doubly-fed induction generators.
	Variable-speed with permanent magnet generators and full-rating frequency converters.
Enercon	Variable-speed with multi-pole synchronous generators and full-rating frequency converters.
Gamesa Eolica	Variable-speed with doubly-fed induction generators.
Siemens	Fixed-speed with conventional induction generators.
	Variable-speed with induction generators and full-rating frequency converters.
	Variable-speed with permanent magnet generators and full-rating frequency converters.
RePower	Variable-speed with doubly-fed induction generators.
Nordex	Fixed-speed with conventional induction generators.
	Variable-speed with doubly-fed induction generators
Others	Fixed-speed with conventional induction generators.
	Variable-speed with doubly-fed induction generators.

Table 1.2   Largest wind turbine manufacturers and their main products. Source: wind turbine manufacturers.

Country	Present and planned wind power incorporation
Spain	10 GW in the year 2005 and up to 12 GW by the year 2010.
Germany	18.4 GW in the year 2005 and up to 30 GW by the year the year 2030.
Great Britain	1.1 GW in the year 2005 and expected 6.2 GW by the year 2010.
The Netherlands	1.2 GW in the year 2005 and announced 6 GW by the year 2020.
Sweden	500 MW in the year 2005 and planned 2.7 GW offshore by the year 2015.
USA	6.4 GW and as prognosis between 10 and 84 GW depending on legislation.
Australia	250 MW and proposed 6.7 GW.

Table 1.3   Selected figures for global wind power incorporation. Sources: European Wind Energy Association, British Wind Energy Association, American Wind Energy Association, Australian Wind Energy Association.

Figure 1.2   Manufacturing wind turbines - snap-shots of the American manufacturer GE Wind Energy. Photo copyright GE Wind Energy. Reproduced with permission from GE Wind Energy.

Figure 1.3 Assembling of the 3.6 GE wind turbines at the Arklow offshore windfarm, Ireland and view of the Arklow offshore windfarm. Photo copyright GE Wind Energy. Reproduced with permission from GE Wind Energy.

However, let us keep in mind the figures of **Table 1.3** will never be complete because large new projects are announced world-wide practically said every week. The readers are therefore addressed to the news updated at websites of the wind turbine manufacturers and the wind energy associations.

1.3 Power system stability

When the amount of wind power increases, power generation from conventional power plants must be reduced to balance power generation and consumption. Presently, conventional power plants are kept in operation to provide voltage and frequency control throughout the grid. The amount of wind power commissioned in Western Denmark is however so significant that the situations where power generated by wind turbines alone can meet all the demand have been observed (Akhmatov, 2003(a)). **Figure 1.4** provides an example on this.

Such situations may introduce a challenge with regard to controlling the voltage and balancing the power grid, thus affecting system stability. **Figure 1.5** presents the general classification of power system stability.

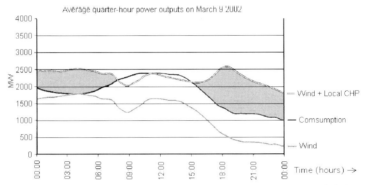

Figure 1.4   Generation from wind and local CHP versus consumption in Western Denmark during March 9 2002. Shaded area is when demand is less than supply from wind and CHP. Courtesy of Energinet.dk, Transmission System Operator of Denmark.

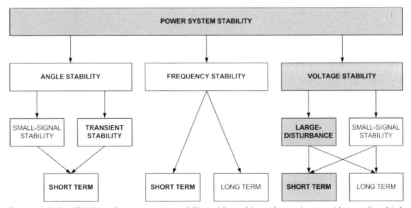

Figure 1.5   General classification of power system stability with marking relevant issues with regard to this book.

Dynamic wind turbine models to be described in this book relate first and foremost to the issue of voltage stability at a short-circuit fault in the grid. As a short-circuit fault may result in significant voltage drop, then this modelling work relates to large-disturbance stability. The duration of such a short-circuit faults can be in the range of hundreds of milliseconds and the total duration of the simulation is in the range of seconds or of tens of seconds. Therefore, this modelling work focuses upon short-term stability issues. Therefore, the main issue of wind turbine modelling in this book relates to short-term, large-disturbance, voltage stability which is denoted as short-term voltage stability. The blocks relating to this kind of power system stability are shaded in **Figure 1.5**.

The wind turbine models to be described in this book may also be applied to investigations of (i) short-term frequency stability because they contain the control of active power and (ii) short-term (transient) angle stability when the power systems investigated are with synchronous generators used in conventional power plants. The relevant blocks in **Figure 1.5** are marked by bold font.

Furthermore, dynamic wind turbine models of this book can easily be adapted to investigations of small-signal stability by addition of small-signal disturbance sources and, at the same time, by simplification of selected parts of the dynamic wind turbine model. Small-signal stability sources may be present in the power grid itself or stem from dynamic wind fluctuations, blade-tower-passage phenomenon or other mechanical oscillatory sources in the wind turbine construction.

Note that investigations of long-term stability may require other kinds of dynamic models of wind turbines, including representation of wind dynamics.

1.4   Grid Codes for windfarms

Since commissioning of wind turbines in Denmark started in the 1970s, the country has a unique situation due to the major growth in wind power in the Danish power grid is caused by establishment of many local wind turbine clusters. Such wind turbine clusters are scattered mainly throughout Western Denmark and in the islands of Lolland and Falster in Eastern Denmark, and are connected to the local distribution grids at voltage levels of 60 kV and below. When the transmission power grid is subject to a short-circuit fault, the local wind turbines may disconnect from the power grid. Then the grid voltage will be re-established by the excitation control of conventional power plants. Conventional power plants will then also increase their active power generation to compensate for the power loss caused by the tripping of the local wind turbines and so balance power in the grid.

As the installed power capacity of local wind turbines is significant and still increasing, the Grid Code for the wind turbines connected to the Danish power grids at the voltage levels below 100 kV was put into force in July 2004 (Energinet.dk, 2004(a)). With regard to maintaining short-term voltage stability, the Grid Code focuses on the fault-ride-through capability of the local wind turbines, which means that sub-sequential disconnection of wind turbines at grid faults is not accepted.

According to the Danish Grid Code (Energinet.dk, 2004(a)), wind turbines must ride through short-circuit faults when the voltage profile in the common connection point of the cluster is as shown in **Figure 1.6**. As older wind turbines installed in the 1970s and the 1980s will be replaced eventually by new and more efficient wind turbines, these new wind turbines are now subject to the Grid Code (Eenerginet.dk, 2004(a)).

In years to come, the major increase in wind power will be gained by incorporation of large (offshore) windfarms. The installed power capacity of such windfarms is so large that they are directly connected to the transmission power grids which are operated at voltages above 100 kV. The Danish offshore windfarm at Horns Rev (shown in **Figure 1.7**) is an example on such a large windfarm connected to the Danish transmission power grid at a voltage level of 165 kV.

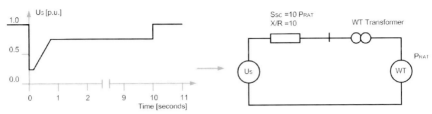

Figure 1.6 Standardised voltage profile to test the fault-ride-through capability of wind turbines to be commissioned in the Danish power grid. WT denotes the wind turbine generator, P_{RAT} is rated power, U_S is voltage, X/R is ration of reactance to resistance of the line, S_{SC} is short-circuit capacity. Courtesy of Energinet.dk, Transmission System Operator of Denmark.

Figure 1.7 The Horns Rev offshore windfarm under construction in summer 2002. Photo copyright Vestas Wind Systems. Reproduced with permission from Vestas Wind Systems.

The Danish Grid Code for wind turbines connected to the power grid at voltages above 100 kV was put into force in December 2004 (Energinet.dk, 2004(b)) and has replaced the code of (Eltra, 2000). Notice that the Danish Grid Code (Eltra, 2000) is still valid for large offshore windfarms Horns Rev A and Rødsand 1 / Nysted commissioned before December 2004.

The Grid Code (Energinet.dk, 2004(b)) contains also the requirement on the fault-ride-through capability of wind turbines regarding standardised voltage profiles that in simulations are inputted into the windfarm connection point, see **Figure 1.6**. Wind turbine manufacturers and windfarm owners are obliged to provide the documentation proving that wind turbines will ride through the short-circuit fault resulting in such a voltage profile.

Windfarm owners are also obliged to provide a validated model of the windfarm with all the necessary control and protection to be applied for investigations of short-term voltage stability. As the focus is moving from small local wind turbine sites towards large (offshore) windfarms which must comply with national Grid Codes, the complexity of detail of dynamic wind turbine models to be applied for investigations of short-term voltage stability must contain a sufficient representation of the control and protection which allows the wind turbines to ride through such grid faults. For example, when power electronics converters are used, protective converter sequences may result in reduced power generation or in loss of specific control provided by such converters during normal operation. Such protective converter sequences must be represented in the dynamic wind turbine model as they may introduce a significant impact on operation of the transmission power grid during grid faults.

Other power systems such as those used in Ireland, Great Britain and Australia have avoided the stage of intensive incorporation of electricity-producing wind turbines of older concepts in small clusters and experience incorporation of large (offshore) windfarms. The national system operators in these countries have formulated Grid Codes for the grid-connection of such large windfarms. Such Grid Codes are "living" documents and are continuously updated by the national system operators as the wind technology is constantly improving, larger incorporation of wind power is taking place and larger impact on the power grids is expected. The Grid Codes are available on the websites of the national system operators presented in **Table 1.5**.

System Operator	Web-link
Denmark	www.energinet.dk
E.On Netz GmbH /Germany	www.eon-netz.com
Red Eléctrica de España /Spain	www.ree.es
ESB National Grid /Ireland	www.eirgrid.com
National Grid UK /Great Britain	www.nationalgrid.com/uk
National Grid Code Administrator /Australia	www.neca.com.au

Table 1.5 Links to selected national system operators providing the Grid Codes for connection of wind turbines.

With regard to maintaining the short-term voltage stability, all the Grid Codes require that the voltage in the transmission power grid is re-established without subsequent disconnection of the large (offshore) windfarms - i.e. the fault-ride-through capability.

2   Induction generator based concepts

The major part of wind turbine concepts are today induction generator based. This implies that an induction generator is applied to convert the mechanical power of the rotor into the active electrical power supplied to the power grid. The first wind turbines of such induction generator based concepts have been fixed-speed and equipped with conventional induction generators with a short-circuited rotor circuit. The presence of an electromechanical slip that is the relative difference between the electrical speed of the power grid and the mechanical speed of the generator rotor, provides a relatively flexible coupling between the fluctuating power of the rotor (with a fluctuating rotor speed) and the electrical power grid (with a fixed frequency). The fluctuating behaviour of the rotor mechanical power is caused by the fluctuating nature of wind and also by some effects of the wind turbine structure such as the blade-tower-passage (Akhmatov and Knudsen, 1999).

Commercial production of fixed-speed wind turbines equipped with conventional induction generators started in Denmark in the early 1970s. The rated power of the first commercial wind turbines was in the range of 20 kW. But by the 1990s, the rated power capacity of wind turbines reached and exceeded 1 MW. At this time, blade-angle control and power electronics converters entered wind technology and the wind turbine market.

The use of advanced control systems implies that the induction generator itself can be used together with different control systems. The control applied to induction generators may be set to reduce flicker emission, improve the efficiency and the power factor, and control the reactive power. Although the wind turbines are still induction generator based, the characteristics of the wind turbines using power electronics converters may be changed drastically when compared to the conventional fixed-speed wind turbines. The use of power electronics converters together with induction generators has become a topical issue in the control of modern wind turbines. Most wind turbines above 2 MW are now variable-speed wind turbines where the variable-speed operation is gained by application of power electronics converters.

The manufacturing of wind turbines has become a large, rapidly-developing business penetrated by advanced wind technology, mechanics and power electronics converter technology. **Figure 2.1** shows a schematic of a 3.6 MW variable-speed wind turbine equipped with a doubly-fed induction generator from the American manufacturer GE Wind Energy. This wind turbine is designed by GE Wind Energy for offshore applications. Such advanced variable-speed wind turbines represent advances in wind technology that will, in the near future, have similar functionality and controllability compared to conventional power plants. Here we can mention the fault-ride-through capability of modern wind turbines, the ability to control reactive power control and support grid voltage, the frequency control etc. **Figure 2.2** shows the offshore project of GE Wind Energy in Arklow, Ireland, with 7 turbines of the 3.6 MW class. **Figure 2.3** presents the Horns Rev offshore windfarm, Denmark, commissioned with eighty 2 MW wind turbines from Vestas Wind Systems. These projects are among the first windfarms ahead to wind power plants.

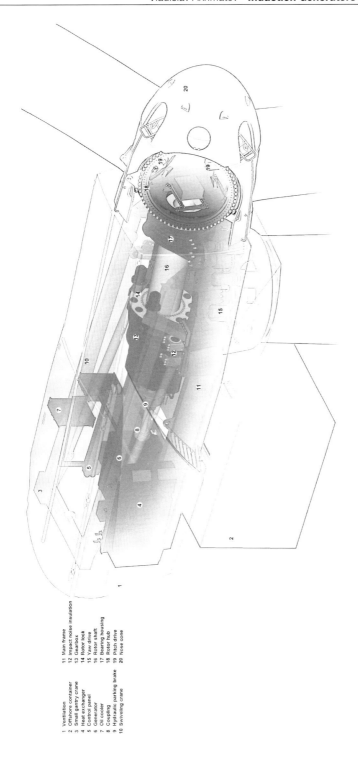

1 Ventilation
2 Offshore container
3 Small gantry crane
4 Heat exchanger
5 Control panel
6 Generator
7 Oil cooler
8 Coupling
9 Hydraulic parking brake
10 Swiveling crane

11 Main frame
12 Impact noise insulation
13 Gearbox
14 Rotor lock
15 Yaw drive
16 Rotor shaft
17 Bearing housing
18 Rotor hub
19 Pitch drive
20 Nose cone

Figure 2.2 The Arkow offshore windfarm containing seven 3.6 MW wind turbines from GE Wind Energy. Photo copyright GE Wind Energy. Reproduced with permission from GE Wind Energy.

Figure 2.3 Commissioning of the Horns Rev offshore windfarm, summer 2002. Photo copyright Vestas Wind Systems. Reproduced with permission from Vestas Wind Systems.

Fixed-speed wind turbines applied together with auxiliary equipment such as dynamic reactive compensation units are also suitable for large windfarms. The use of blade-angle control and dynamic reactive compensation contributes to the fault-ride-through capability of such fixed-speed wind turbines and can improve their control characteristics (Akhmatov et al., 2003(a)). The Rødsand 1 / Nysted offshore windfarm commissioned in Denmark in 2003, contains seventy-two Siemens 2.3 MW units and is at present the largest offshore windfarm with fixed-speed, active-stall controlled wind turbines in the world.

Despite differences in the induction generator-based concepts, the main components of a modern wind turbine with regard to their impact on the power grid are the induction generator itself, the shaft system, the three-bladed rotor with blade-angle control, the power electronics converter (if any applied) and the protective relay system. A short conceptual survey below presents existing wind turbine concepts where induction generators are applied to convert the mechanical power of the three-bladed rotor into the active power supplied to the grid.

2.1 Conventional fixed-speed concept

The oldest commercial wind turbine concept uses induction generators with a short-circuited rotor circuit. In this concept, the generator rotor is coupled to the wind turbine rotor through a shaft system, whereas the stator circuit is AC- connected to the power grid, see **Figure 2.4**. This is termed the fixed-speed concept because the speed in normal operation may only vary within a narrow range that is normally up to 2%. This speed range is defined by the electromechanical slip of the induction generator.

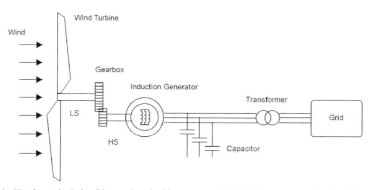

Figure 2.4 Fixed-speed wind turbine equipped with compensated induction generator. Reprinted from Ref. (Akhmatov, 2003(b)), Copyright (2003), with permission from the copyright holder.

The concept of fixed-speed wind turbines is also termed the Danish concept. Presently, most wind turbines commissioned in Denmark are such fixed-speed turbines equipped with induction generators. **Figure 2.5** shows the Middelgrund offshore windfarm commissioned near to the Danish capital Copenhagen. The Middelgrund windfarm consists of twenty Siemens 2 MW fixed-speed, active-stall controlled wind turbines.

Fixed-speed wind turbines are also commissioned in numerous sites in Europe and America. **Figure 2.6** shows Siemens fixed-speed wind turbines at Galicia, Spain.

Figure 2.5 The Middelgrund offshore windfarm located near Copenhagen. Photo copyright Siemens. Reproduced with permission from Siemens WPG.

Figure 2.6 Fixed-speed wind turbines located in the site Galicia, Spain. Turbine type: Siemens 600 kW. Project size: 39.6 MW. Photo copyright Siemens. Reproduced with permission from Siemens WPG.

Induction generators supply active power to the grid, but absorb reactive power from the grid. The reactive power absorption is required to excite the generator. Obviously, induction generators with a short-circuited rotor circuit cannot control the reactive power and require that the grid voltage is kept about the generator rated voltage. Often, the fixed-speed wind turbines equipped with induction generators are either no-load compensated or fully compensated with the use of capacitor banks. Such a compensation arrangement is applied to reduce the total reactive power absorption from the power grid and to improve the power factor of the wind turbine.

The shaft system contains the low-speed shaft (LS) connected to the three-bladed rotor, the high-speed shaft (HS) connected to the generator rotor and the gearbox. The gearbox is required to provide transformation between the slow-rotating rotor and the fast-rotating generator rotor. The synchronous rotor speed of modern fixed-speed wind turbines may be in the range 15 to 20 rev./min. The synchronous speed of the generator rotor is coupled to the grid frequency and 1500 rev./min. for generators with two pole-pairs and 1000 rev./min. for generators with three pole-pairs when the rated grid frequency is 50 Hz as in European countries and Australia. The mechanical gear ratio of modern wind turbines may then approach 100.

Fixed-speed wind turbines are either fixed-pitch or with blade-angle control. Fixed-pitch wind turbines are often called stall-controlled because their mechanical power is limited and controlled at the rated power during rated wind conditions due to the stall effect. This "passive" stall control is slow, with the blades simply bolted onto the hub at a fixed pitch angle. At a given wind speed, the wind turbine rotor starts to stall. Stall conditions start at the blade root and develop gradually across the whole blade length as the wind speed increases. The main advantage of this design is that this is a robust and cheap solution. However drawbacks of stall control are the relatively low efficiency at low wind speed, variations in the maximum steady-state power caused by variations in the air density and uncontrollable emission of flicker.

Blade-angle control is applied in modern fixed-speed wind turbines to improve efficiency and to eliminate variations in the maximum steady-state power due to variations in air density. The blade-angle control of fixed-speed wind turbines is usually active-stall which applies operation in the negative range of pitch angles. The active-stall control is set to optimise power output in wind speeds below the rated wind and to keep the power output at the rated level (power limiting mode) when the wind speed exceeds the rated level. In the power limiting mode, the blades are turned into a deeper stall region by increasing the angle of attack. The active-stall control is characterised by a smooth limitation of power when the wind speed exceeds the rated wind. The active-stall control contributes to reduction of flicker emission from the wind turbines.

2.2 Concept with dynamic rotor resistance

Figure 2.7 presents a conceptual sketch of partly-variable-speed wind turbines equipped with induction generators with dynamic rotor resistance (DRR). In this concept, the rotor circuit is connected to the power electronics converter. Operation of the power electronics converter is controlled by the Insulated Gate Bipolar Transistor (IGBT) -switches and in this concept corresponds to adding an external resistance in series to the rotor circuit. This concept corresponds to OptiSlip® wind turbines produced by the manufacturer Vestas Wind Systems.

The dynamic control of the external rotor resistance allows continuous operation of the generator rotor slip. The slip range is from that of a conventional induction generator with a short-circuited rotor circuit (about 2% at rated operation) to 10%. Such slip variation allows operation at partly variable-speed in the super-synchronous range up to 10% above the synchronous speed. This control is mostly applied to reduce flicker emission from the wind turbines to the power grid as the mechanical power fluctuations are converted to the kinetic energy of the rotor and absorbed by the external rotor resistance of the converter.

These partly-variable-speed wind turbines are equipped with pitch control. Pitch control is applied to optimise the power output from the rotor in wind speeds below the rated wind and to keep at the rated power in wind speeds above the rated wind. When the pitch control is applied, the blades are turned from the wind when power output becomes excessive. This reduces the angle of attack and limits the mechanical power of the rotor. The use of the pitch control compensates for variations in the maximum steady-state power caused by variations in air density.

Fast pitching rates may however excite overshoots in the mechanical power of the rotor (Øye, 1986). In this case, the power electronics converter minimises occurrence of such overshoots in the active power of the induction generator. Then, potential power overshoots are put into the kinetic energy of the rotor and absorbed by the external rotor resistance. The reaction time of the converter control is significantly smaller than that of pitch control.

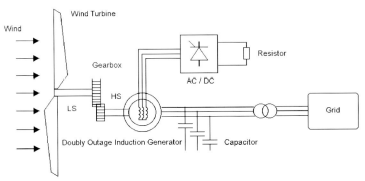

Figure 2.7 Partly-variable-speed, pitch controlled wind turbine equipped with compensated doubly-outage induction generator with dynamic rotor resistance. Reprinted from Ref. (Akhmatov, 2003(b)), Copyright (2003), with permission from the copyright holder.

The control of the power electronics converter does not influence the excitation of the induction generator. The induction generator absorbs the reactive power from the grid analogously to the case of fixed-speed wind turbines. The induction generator is compensated with the use of the capacitor bank to improve the power factor of the wind turbines.

OptiSlip[R] wind turbines produced by Vestas Wind Systems are denoted V47 and V80 (Vestas, 2003). The rated power of the V47 wind turbine is 660 kW. The V47 is the first generation of OptiSlip[R] wind turbines. **Figure 6.1** shows a cluster of OptiSlip[R] V47 wind turbines commissioned in Texas, USA. The rated power of the V80 wind turbine is 1.8 MW and this is the second generation of OptiSlip[R] wind turbines from the manufacturer Vestas Wind Systems. The second generation of the OptiSlip[R] system is basically the same system as for the first generation. The most important change is the introduction of slip-rings and the application of a water-cooled external resistance unit (Vestas, 2003).

2.3   Doubly-fed induction generators

In the early 1990s, the manufacturer Vestas Wind Systems offered the V52 wind turbine as a variable-speed, pitch controlled wind turbine equipped with a doubly-fed induction generator (DFIG) with a partial-load frequency converter. The rated power of the V52 wind turbine is 850 kW. **Figure 2.8** shows a schematic of a variable-speed, pitch controlled wind turbine equipped with DFIG. This wind turbine concept has become very popular. All large wind turbine manufacturers offer such variable-speed wind turbines with DFIG, with exception of Enercon and Siemens. The range of rated power of such variable-speed wind turbines with DFIG available on the market is from 850 kW to 3.6 MW. **Figure 2.9** shows the V90 wind turbine with a rated power of 3 MW from Vestas Wind Systems. Prototypes with rated powers of up to 5 MW have been erected and will enter the market soon.

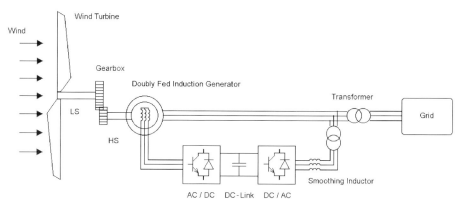

Figure 2.8   Variable-speed, pitch controlled wind turbine equipped with a doubly-fed induction generator with a partial-load frequency converter. Reprinted from Ref. (Akhmatov, 2003(b)), Copyright (2003), with permission from the copyright holder.

Figure 2.9   V90 3 MW wind turbine from Vestas Wind System erected in Nakskov, Denmark. Photo copyright Vestas Wind Systems. Reproduced with permission from Vestas Wind Systems.

In the DFIG based concept, the slow-rotating rotor is connected to a fast-rotating generator rotor through a geared shaft system with a gear ratio of up to 100. The generator stator is AC- connected to the power grid in a similar way to the above-discussed wind turbine concepts. The generator rotor is connected to the power grid through an AC/DC/AC frequency converter which distinguishes this concept from others.

The frequency converter consists of two back-to-back voltage-sourced converters (VSC) connected through a DC- link. The rotor VSC is connected to the rotor circuit through sliprings. The grid-side VSC feeds into the power grid via a smoothing inductor and a transformer. The back-to-back VSC are controlled by IGBT-switches. The frequency converter is required to provide electrical coupling between the rotor circuit which operated at varying frequency at up to 10 Hz to the power grid characterised by a fixed frequency of 50 Hz in Europe and Australia and 60 Hz in the USA. The frequency converter also serves other important purposes.

The rotor VSC induces the voltage vector in the rotor circuit that then has a suitable magnitude and rotates with a desired (variable) frequency. The generator rotor and the wind turbine rotor do not require operation at a fixed speed, but rotor speed can be adjusted by the dynamic control of the rotor VSC. This makes it possible to operate such wind turbines within a wider speed range. Thus these wind turbines are denoted as variable-speed. Variable-speed wind turbines with DFIG made by Vestas Wind Systems are termed OptiSpeed[®3] and operate in the relative speed range from -40% to 15% (and dynamically up to 30%) compared to the synchronous speed.

Accordingly to Vestas (2001), OptiSpeed[®] wind turbines increase energy production by 5% compared to conventional fixed-speed concept. This is because wider speed ranges offer better optimisation of the power output in wind speeds below the rated level. In such an optimisation mode, the rotor converter causes the generator rotor speed to follow the optimised speed, optimised to gain the maximum power efficiency at the given wind speed, see **Figure 3.5**. In light winds, the wind turbine rotor is slow-rotating (sub-synchronous range of speeds) whereas, in strong winds, the rotor rotates faster than the synchronous speed (super-synchronous range of speeds) which increases the power output of the rotor.

The rotor VSC control allows power fluctuations caused by wind fluctuations and gusts to be converted into the kinetic energy of the rotor and smoothly converted into the electrical energy supplied to the power grid. This reduces the impact on the gearbox and the flicker emission to the grid, and also improves the efficiency of the wind turbine with regard to its energy output.

Such variable-speed wind turbines are pitch controlled. Pitch control is applied to optimise the power output in wind speeds below the rated wind and to keep the rated power at wind speeds above the rated wind. The reaction time of the mechanical pitch control is much larger than the reaction time of the frequency converter. The pitch control may even become a delaying factor in relation to optimal power production (Vestas, 2001). In this case, the rotor VSC is set to rapid adjustment of rotor speed to the optimised value, giving the pitch control system time to turn the blades into the desired position. Potential overshoots introduced by the pitch control at fast pitching rates may be put into the kinetic energy of the rotor and smoothly converted into active power.

Rotor VSC control is arranged with independent control of active and reactive power. This implies that the generator is not excited from the power grid. The generator is excited from the rotor circuit with the use of the VSC control. This also means that the generator can be set to control the reactive power and support the grid voltage in undisturbed operation of the power grid. There is therefore no need for capacitor banks for compensation of such wind turbines equipped with DFIG.

---

[3] OptiSpeed[®] is a registered trade mark of Vestas Wind Systems.

Different wind turbine manufacturers have announced specific schemes for reactive power control. For example, the manufacturer GE Wind Energy applies a patented solution called Wind-VAR^{TM4} that is available in all GE Wind Energy megawatt-class wind turbines and eliminates undesired voltage increases and fluctuations. With the use of WindVAR^{TM}, voltage regulation is delivered to the power grid in a fraction of a second. This is especially advantageous in weak power grids or in installations which are located at a distance from the point of delivery. **Figure 2.10** shows GE Wind Energy 1.5 MW wind turbines equipped with WindVAR^{TM} control, erected in Gatun, Spain.

Figure 2.10   Thirty-three 1.5 MW wind turbines from GE Wind Energy erected in Gatun, Spain. Photo copyright GE Wind Energy. Reproduced with permission from GE Wind Energy.

The frequency converter rating is only a fraction of the wind turbine rated power, e.g. this variable-speed concept applies smaller converters. The frequency converter rating is slightly above the generator rated power multiplied by the rated generator slip and typically in the range of 25% of the wind turbine rated power.

Note that the slip is positive in sub-synchronous operation and negative in super-synchronous operation of the rotor. Active power of the rotor circuit is approximately a product of the generator shaft power and the slip. Active power is then supplied from the rotor circuit to the power grid in super-synchronous operation, whereas the active power is absorbed by the rotor circuit from the power grid in sub-synchronous operation. The active power of the stator is always supplied to the power grid, independent of speed. **Figure 2.11** illustrates the active and the reactive power flow in a DFIG.

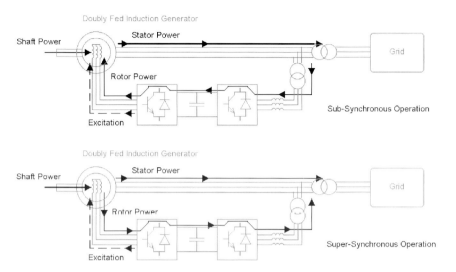

Figure 2.11   Active and reactive power flow in a doubly-fed induction generator with regard to speed regime. Notice that the generator is excited from the rotor VSC and does not exchange reactive power with the grid.

The grid-side VSC is set to control the DC- link voltage and is usually kept reactive neutral, e.g. does not exchange reactive power with the grid in normal operation. In emergency situations, it may be advantageous to set this VSC to contribute to reactive power control.

2.4   Induction generators with full-rating converters

Induction generators may be connected to the power grid through full-rating AC/DC/AC power electronics converters. Before 2005, such wind turbines were only available with power ratings in the range of hundreds kW (Wang and Chang, 2001). In 2005, the manufacturer Siemens announced 2.3 MW and 3.6 MW wind turbines of this concept, but technical details of the wind turbines have so far been kept confidential. **Figure 2.12** presents an outline of this concept (Akhmatov, 2005(a)).

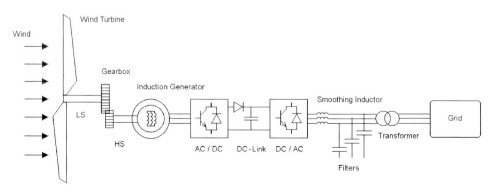

Figure 2.12   Conceptual sketch of variable-speed wind turbines with converter-connected induction generators.

Such wind turbines are variable-speed and so the electrical frequency of the network on the generator-side VSC must follow the optimised synchronous speed of the rotor. The electric frequency of the generator, f_E, is proportional to the wind turbine rotor speed, ω_M, and set by the generator-side converter according to the reference power of the rotor, P_{REF}, (Miller et al., 1997; Wang and Chang, 2001). The variable-speed operation of the wind turbine rotor and the variable-frequency operation of the induction generator are applied to optimise the active power output of the wind turbine.

Figure 2.13 illustrates the optimised relation between the electric frequency set point and the reference power of the wind turbine rotor. Within the optimised operation range, e.g. when the electric frequency is below the rated level, $P_{REF} = k_f \cdot f_E^3$. Here k_f denotes the design factor. The rated electric frequency of the generator is a wind turbine design parameter and can be set in the range of between 10 and 20 Hz. The rated electric frequency is reached when the power output is about 30% of the wind turbine rating. When the power output exceeds this level, the electric frequency is kept at the rated value. Operation of the induction generator is characterised by the electromechanical slip defined with regard to this electrical frequency generated at the induction generator terminals by the generator-side VSC.

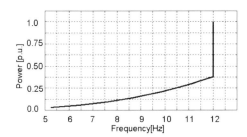

Figure 2.13 Optimised relation between the generator electric frequency and the wind turbine power output. In this example, the rated frequency is 12 Hz.

This VSC also provides excitation of the induction generator. **Figure 2.14** shows the load flow of the converter-connected induction generator that is the active power and the reactive power (excitation of the generator) in steady-state. The induction generator supplies active power to the grid through a back-to-back converter system. The induction generator is de-coupled from the AC power grid through a DC-link which is why the generator cannot be excited from the AC power grid. The induction generator is then excited from the generator-side converter as illustrated in **Figure 2.14.a**. This control arrangement requires that the generator-side converter rating is increased to handle the active power, as well as the reactive absorption of the induction generator. For example, when the generator is at its rated operation, $P_G = P_{RAT}$, and absorbs $Q_G = -Q_{RAT}$, then the converter rating must cover the rated apparent power of the induction generator, $S_{RAT} = \sqrt{P_{RAT}^2 + Q_{RAT}^2}$, as well as the losses in the converter.

Application of fixed or switched capacitors connected to the induction generator terminals (as shown in **Figure 2.14.b** and **Figure 2.14.c**) reduces the demand for reactive power generation by the generator-side converter. This also reduces the generator-side converter rating, because this must handle less reactive power. If the capacitors generate Q_{FC}, the converter rating can be set to $S_{RAT} = \sqrt{P_{RAT}^2 + \left(Q_{RAT} - Q_{FC}\right)^2}$, plus the losses in the converter.

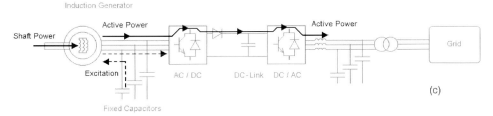

Figure 2.14 Active and reactive power flow in a converter connected induction generator: **(a)** induction generator excited by the generator-side converter, **(b)** induction generator at high-load operation excited in part by the fixed capacitors and by the generator-side converter, **(c)** induction generator at low-load operation excited by fixed capacitors, and the generator-side converter may absorb reactive power surpluses.

Note that the induction generator absorbs more reactive power when being in high-load operation, e.g. in strong winds. Thus, the induction generator will, in part, be excited by the generator-side converter and by the fixed capacitors. This high-load operation situation is illustrated in **Figure 2.14.b**. **Figure 2.14.c** shows the low-load operation which occurs in light winds. Here the induction generator absorbs less reactive power and the generator-side converter can be set to absorb the surplus of the reactive power from the fixed capacitors, preventing over-excitation of the induction generator. A surplus of reactive power may be present when the reactive power demands of the generator, Q_G, are less than the reactive power generated by the fixed capacitors, Q_{FC}.

The shaft system again consists of a low-speed from the rotor, and a high-speed shaft entering the generator, coupled via a gearbox. The gearbox ratio is however reduced in this concept because the electric frequency at the induction generator terminals is lower than that of the grid.

The grid-side converter is set to control the DC-link voltage to balance the power flow from the induction generator to the power grid. The grid-side VSC is also set to control the reactive power and improve the wind turbine power factor. In emergency situations, this converter may be set to support the grid voltage. Such variable-speed wind turbines are equipped with blade-angle control (either pitch or active-stall) for better power output optimisation in light winds and to keep the rated power output in strong winds. At the terminals of grid-side VSC, there are a smoothing reactor and filters to form a sinusoidal phase voltage and attenuate higher harmonic emission from the VSC switches.

2.5 Representation in stability studies

The simulation tool and the level of the models' complexity must be in agreement with the purpose of investigations. Furthermore, the models to be applied in investigations must be in agreement with assumptions of the chosen simulation tool, e.g. the models may not cause conflicts with numerical algorithms of the simulation tool. When using user-written models, these must comply with the simulation tool interface, e.g. the roles for exchanging data and results between the model and the simulation tool.

Most investigations of short-term voltage stability relate to situations where the transmission power grid is subject to a 3-phase short-circuit fault. This kind of a fault is a balanced symmetrical event, i.e. the transmission power grid and the induction generators can be represented by their positive-sequence equivalents. The negative- and zero-sequence equivalents may be excluded from the power grid model as well as from the induction generator model. When the stability and fault-ride-through capability of wind turbines are evaluated with regard to asymmetrical short-circuit faults, then representation of the power grid and of the induction generator must contain their positive-sequence equivalents as well as the negative- and zero-sequence equivalents. The Danish Grid Code (Energinet.dk, 2004(b)) requires that the wind turbines must ride through 3-phase short-circuit faults that are balanced symmetrical events as well as through phase-to-phase faults and single-phase short-circuit faults that are asymmetrical events.

Below we provide a short list of simulation tools applied to investigations of short-term voltage stability and wind turbine modelling.

With the use of the simulation tool PSS/E[TM5] investigations of short-term voltage stability can be carried out with regard to single-phase and 3-phase short-circuit faults in the power grid. The simulation tool Powerfactory[TM6] allows simulations of 3-phase short-circuit faults as well as any kind of asymmetrical faults in the power grid. These two simulation tools also provide coupling to the simulation tool Matlab/Simulink[®7] allowing application of the models of the tool Matlab/Simulink[®] in simulations carried out using PSS/E[TM] or Powerfactory[TM].

The simulation tool SimPow[®8] can be used to model generators as well as power electronics components and is suited to the complex modelling of power electronics devices such as converters and compensation equipment. SimPow[®] however allows simulations of 3-phase short-circuit faults as well as any kind of asymmetrical faults.

---

[5] PSS/E[TM] is a trade mark of Siemens Power Technologies, Inc. Schenectady, N.Y., USA.

[6] Powerfactory[TM] is a trade mark of DigSilent GmbH, Gomaringen, Germany.

[7] Matlab[®] and Simulink[®] are registered trade marks of The MathWork, Inc.

[8] SimPow[®] is a registered trade mark of STRI AB, Sweden.

The major simulation tools applied to investigations of short-term voltage stability and fault-ride-through capability contain models of induction generators. The tool Matlab/Simulink® contains a 3-phase model of induction generators in the power-set block as well. The next step in developing wind turbine models is the creation of relevant models of wind turbine mechanics, aerodynamics and control. Block-diagrams for different wind turbine concepts with the use of induction generators are shown in **Figures 2.15** to **2.18** and discussed in the following sections. Such block-diagrams show the mechanical and aerodynamic components of the wind turbines, their control, protection and interaction with the power grid.

All the above-mentioned simulation tools also have reduced student-versions that are offered by the tool suppliers to educational programmes and universities.

2.5.1   Fixed-speed wind turbines

Figure 2.15 presents a generic block-diagram of a fixed-speed wind turbine equipped with a conventional induction generator with a short-circuited rotor circuit. Each block of the model represents a physical component of the wind turbine, and the arrows give links between the different model blocks. Such links show the data exchange between the component blocks within the dynamic wind turbine model, and the power flow between the generator and the power grid. The main components of this model are:

1) Dynamic wind (if any wind fluctuations are required in the investigations).
2) Aerodynamic rotor.
3) Shaft system.
4) Induction generator and its capacitor bank.
5) Protective system.
6) Blade-angle control (active-stall control) if any.

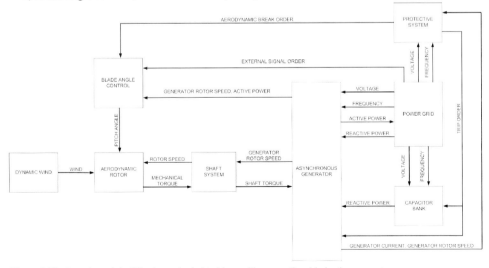

Figure 2.15   Generic model of fixed-speed wind turbines with conventional induction generators.

Fixed-speed wind turbines are a relevant concept for large (offshore) windfarms. Therefore such fixed-speed wind turbines may be subject to the Grid Code of the local system operator. For example, the Danish Grid Codes (Eltra, 2000; Energinet.dk, 2004(b)) require that the active power supply from a large windfarm to the transmission power grid may be reduced by an external order sent to the large windfarm from the external system. The reactive power control, such as the capacitor bank of the large windfarm, must be available for the external system to support voltage in the case of grid disturbances. The external system is defined as the power grid outside the windfarm and refers to the system operator. In such conditions, the external system controlling the power set point of the wind turbines and their capacitor banks must be part of the dynamic model of fixed-speed wind turbines. In **Figure 2.15**, this optional control is indicated by arrows going from the external system to the blade-angle control so as to control the active power set point, and to the capacitor bank to control the reactive power exchange between the wind turbines and the power grid.

2.5.2   Wind turbines with a dynamic rotor resistor

Generic block-diagram model of pitch controlled wind turbines equipped with induction generators with a dynamic rotor resistance (DRR), often denoted as variable-slip generators, is shown in **Figure 2.16**. Besides the components which are present in fixed-speed wind turbines, this model contains a rotor converter system. The rotor converter is connected to the rotor circuit of the generator, the converter operates then as the external resistance in series with the rotor circuit, continuously regulated by the rotor converter control mechanism. The converter control receives a signal for the desired power, monitors the rotor current and adjusts the external resistance accordingly to minimise fluctuations in rotor current magnitude. The model contains the following blocks:

1) Dynamic wind (if any wind fluctuations are required in the investigations).
2) Aerodynamic rotor.
3) Shaft system.
4) Induction generator and its capacitor bank. The rotor circuit is with an external rotor resistance adjusted by the rotor converter.
5) Rotor converter including control and interfacing the generator model of the simulation tool.
6) Protective system.
7) Pitch control.

When such wind turbines are commissioned in large windfarms, the main issue is the fault-ride-through capability of the wind turbines. The rotor converter is controlled by IGBT-switches that are sensitive power electronics devices. To gain the fault-ride-through operation, the solution with blocking and then restarting the power electronics converter is suggested. When the converter blocks, e.g. protects itself, the IGBT-switches stop switching and the rotor circuit is closed through an external rotor resistor which often is called a crow-bar. This solution is therefore denoted as crow-bar protection using the external rotor resistor as the crow-bar. When the rotor converter restarts, the crow-bar is removed, the IGBT-switches start switching again and control the DRR. When required, this solution allowing the wind turbines to ride through the grid faults must be implemented into the dynamic wind turbine model as shown in **Figure 2.16**.

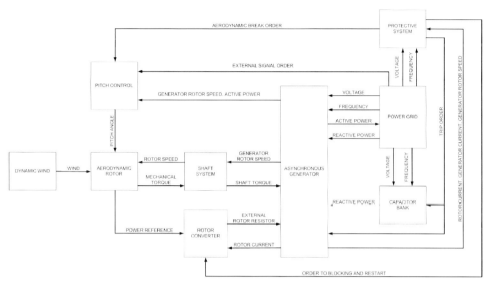

Figure 2.16   Generic model of pitch controlled wind turbines equipped with induction generators with a dynamic rotor resistance.

Additionally, control of the external system may be required accordingly to the national Grid Codes.

2.5.3   Wind turbines with doubly-fed induction generators

Figure 2.17 shows a generic model of variable-speed, pitch controlled wind turbines equipped with DFIG and partial-load frequency converters. The generic model must contain the blocks:

1) Dynamic wind (if any wind fluctuations are required in the investigations).
2) Aerodynamic rotor.
3) Shaft system.
4) Induction generator with accessible rotor circuit. In the rotor circuit, the rotor VSC induces and controls the rotor voltage of a given magnitude and a given phase-angle.
5) Rotor VSC including control and interfacing the simulation tool.
6) Grid-side VSC including control and interfacing the simulation tool.
7) DC link.
8) Protective system.
9) Pitch control.

Note that this model contains a representation of a rotor VSC controlling the active power of the generator and also providing excitation of the generator through the rotor circuit. The model also represents a grid-side VSC providing the active power exchange between the rotor circuit and the power grid, and a DC- link connecting the two VSC to each other.

The model focuses on the fault-ride-through capability of such wind turbines when commis-sioned in large (offshore) windfarms. As the frequency converter contains the IGBT-switches that must be protected against excessive overloads, the dynamic wind turbine model must also contain a converter protection model with the converter blocking and restarting. This control feature can be arranged with the use of crow-bar protection, e.g. closing the rotor circuit through an external resistor (the crow-bar). At abnormal operation in the power grid, the rotor converter blocks and trips, whereas the rotor circuit is short-circuited through the crow-bar. When the grid operation re-establishes, the crow-bar is removed, and then the rotor converter restarts and starts switching again.

Note that other protective sequences of the rotor converter are also feasible (not only the crow-bar protection). For example, uninterrupted operation at the rotor converter applying a current limiting solution can be feasible and useful to support the grid voltage at grid disturbances (Akhmatov, 2002(b)). Presently, development within the fault ride-through solutions is going into this direction.

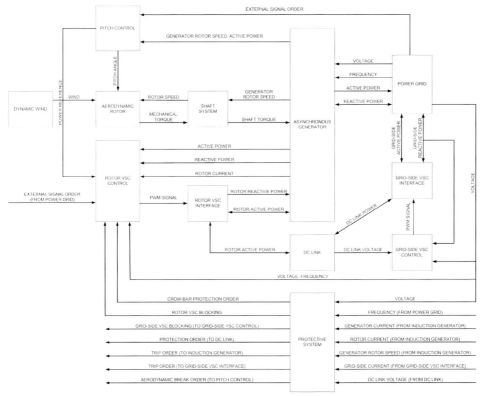

Figure 2.17   A generic model of variable-speed, pitch controlled wind turbines equipped with doubly-fed induction generators and partial-load frequency converters.

Action of the grid-side VSC during this protective sequence must be clarified as well. For example, the grid-side converter may stay in operation and support the grid voltage in the vicinity of the wind turbine terminals.

According to the national Grid Codes, the external system may require the control of active and reactive power provided by the windfarm. In this wind turbine concept, the reactive power set point can be sent to the rotor VSC as well as to the grid-side VSC. The active power set point will be sent to the pitch control systems and then to the rotor VSC as the active power can be rapidly changed by adjusting the generator rotor speed, which is controlled by the rotor VSC. Normally, a power gradient limit is applied on how fast the active power may be changed. This additional control is given by respective blocks and links in the generic block-diagram in **Figure 2.17**.

2.5.4 Wind turbines with induction generators and full-rating converters

This concept is announced by Siemens in 2005 for MW-class wind turbines. **Figure 2.18** shows a generic block-diagram of this concept used for voltage stability investigations. The generic model contains the following blocks:

1) Dynamic wind (if any wind fluctuations are required in the investigations).
2) Aerodynamic rotor.
3) Shaft system.
4) Induction generator operating at fully-variable frequency set by the generator VSC.
5) Generator VSC including control and interfacing the simulation tool.
6) Grid-side VSC including control and interfacing the simulation tool.
7) DC link.
8) Protective system.
9) Pitch control.

When the fault-ride-through capability is required, this must be represented in the converter control and protection. The frequency converter contains IGBT-switches which must be protected against excessive overloads (over-current, over-voltage and thermal impact). Therefore the converter protection model with the converter blocking and restart must be part of the dynamic wind turbine model. This control can be arranged in several ways, i.e.:

1) As a control feature with the generator converter blocking and restart. Here, the converter is ordered to block at excessive over-voltage in the DC- link. It immediately interrupts the power supply to the DC- link from the generator VSC. This may also stabilise the DC- link voltage within an acceptable range. The converter restarts again when normal operation is re-established.
2) Uninterrupted converter operation requiring an immediate frequency change at excessive over-voltage in the DC- link. Here, the power reference of the generator VSC is immediately reduced or even set to a negative value. This results in an almost immediate change of the electric frequency at the generator terminals and interrupts the active power supply to the DC-link. At extremely excessive DC- link voltages, the generator VSC can be set to absorb some power from the DC- link and so reduces the DC- link voltage, whereas the pitch control will handle overspeeding. The power reference and electric frequency are re-established at normal levels when the power grid returns to normal operation.

Additionally, a chopper can be present in the DC- link. The chopper must short-circuit the DC-link capacitor through a resistance at excessive DC- link voltage. In this case, the chopper operates as an electronic crow-bar and must be fast-acting and reliable.

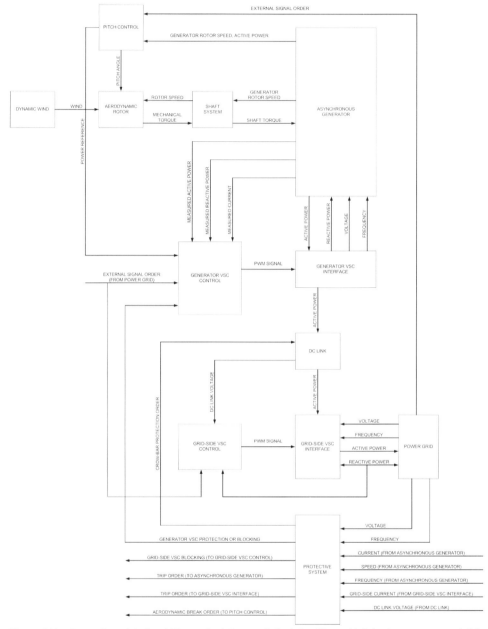

Figure 2.18   A generic model of variable-speed, pitch controlled wind turbines with induction generators and full-rating converters.

The grid-side VSC must be designed to withstand the impact of the grid during such short-circuit faults that may cause blocking of the generator VSC. Obviously, it is advantageous to use the full-rating capability of grid-side VSC to control the reactive power and to support the grid voltage in such short-circuit faults. If the grid-side VSC has to block, this must not cause the wind turbine to stop, and the grid-side VSC must restart very shortly after the grid fault has been removed. Here, experiences from the fault-ride-through design of Static Synchronous Compensators (Statcoms) can be applied to the grid-side VSC of such wind turbines.

When required in the national Grid Codes, the external system may have access to control active and reactive power of the windfarm. In this wind turbine concept, the reactive power set point may be sent to the grid-side VSC of the wind turbines. The active power set point may be sent to the pitch control and then to the generator VSC as the generator VSC provides fast control of the active power by adjusting the frequency at the induction generator terminals. This additional control is given by the respective blocks and links in the generic block-diagram in **Figure 2.18**.

2.6 Summary

Wind turbine manufacturing has become a professional, market-oriented business. The rated power of modern wind turbines has increased 3 MW and is approaching 5 MW. The major commercial wind turbines today are equipped with induction generators.

The conventional induction generators supply active power to the grid and absorb reactive power from the grid. In other words, they are excited from the grid and cannot control voltage. When combined with the control of power electronics converters, the characteristics and the controllability of such induction generators are changed significantly. The converter-controlled induction generators may control the reactive power and offer the grid voltage support. The use of power electronics converters allows variable-speed operation which makes it possible to gain more power from the wind compared to fixed-speed operation. The induction generator based concepts are:

1) Fixed-speed, active-stall controlled wind turbines equipped with conventional induction generators with a short-circuited rotor circuit.
2) Partly variable-speed, pitch controlled wind turbines equipped with induction generators with DRR.
3) Variable-speed, pitch controlled wind turbines equipped with doubly-fed induction generators and partial-load frequency converters.
4) Variable-speed, pitch controlled wind turbines equipped with induction generators and full-rating converters.

Dynamic models of electricity-producing wind turbines must be applied to investigations of short-term voltage stability. Though there are conceptual differences, there are common parts in wind turbine constructions with such induction generators. The common elements include:

1) Aerodynamic rotor.
2) Blade-angle control (pitch or active-stall) if any.
3) Geared haft system.
4) Induction generator. Access to the rotor circuit may be required for specific concepts.
5) Protective system.
6) Converter with control if any.

When the fault-ride-through capability is required, additional control loops for such fault-ride-through are needed. For example, control loops allowing blocking and restart of the power electronics converters are required in the wind turbine concepts of converter-controlled generators. The level of the complexity of each model must be in agreement with the purpose of investigations and the physics of the modelled component.

Note that dynamic wind model can be omitted, as investigations of short-term voltage stability are dedicated to grid events with duration of seconds, during which short period incoming wind may be assumed stationary.

3. Wind turbine construction

A modern wind turbine is a complex mechanical construction. The main parts of a wind turbine are the foundation, the tower, the hub, the three-bladed rotor and the shaft system with a gearbox, see **Figure 3.1(a)** which also illustrates the kinds of problems which are relevant in design and construction stability studies. Among such problems are the optimisation of the mechanical power output, the stability of the foundation relating to offshore sites, the deflection and oscillations of the tower, the deflection, torsion and oscillations of the blade, operation of the rotor in turbulent and irregular winds, torsion of the shaft system, and the forces and moments acting on gearwheels. Many of these technical problems relate to aeroelasticity of the wind turbine construction and can be investigated with the use of aeroelastic codes (Chaviaropoulos, 1996; Hansen et al., 2005).

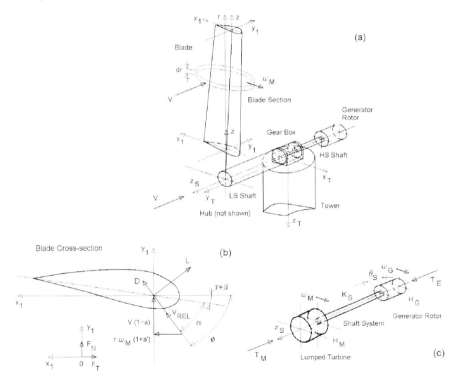

Figure 3.1   Mechanical construction of the wind turbine with regard to the generator operation: **(a)** - Sketch of the mechanical construction, **(b)** - the Blade Element Momentum method applied to a blade section, **(c)** - the shaft system representation with the use of the two-mass model. Reprinted from Ref. (Akhmatov, 2003(c)), Copyright (2003), with permission from Multi-Science Publishing Company.

Such technical tasks to be studied and solved in the design process of a wind turbine construction are useful to know. However such complex models and studies are not necessarily needed when a wind turbine construction must be represented in investigations of power system stability.

Investigations of power system stability focus on the operation of the wind turbine induction generator and the electromechanical interaction between the generator and the power grid. Modelling of the wind turbine construction can be reduced to issues of rotor aerodynamics and shaft system representation.

Rotor aerodynamics are used to determine the mechanical power, P_M, produced by the wind turbine rotor in a given wind. The rotor aerodynamics give a coupling between the rotor speed, ω_M, and the mechanical power produced by the rotor, in stationary wind (grid events with duration of seconds). The rotor aerodynamic provides also a coupling between the pitch angle, β, and the mechanical power of the rotor, P_M.

The shaft system is subject to twist and shows oscillating behaviour at grid disturbances. In this way, the shaft system behaviour influences the induction generator dynamics and represents the electromechanical interaction between the wind turbine and the electric power grid.

This chapter outlines how to perform computations of mechanical power output of a wind turbine, the dynamic behaviour of the rotor and the shaft system, the control of the rotor and understand advantages and restrictions of the methods commonly applied in such computations.

3.1 Basic rotor aerodynamics

At equilibrium, the mechanical torque produced by the rotor and transferred to the low-speed shaft, T_M, and the pressure force (also called the thrust), T, can be computed using the Blade Element Momentum (BEM) method (Freris, 1990; Sørensen and Kock, 1995; Hansen, 2000). According to the BEM method, the rotor blade is separated into a number of equidistant sections along the blade length. For each blade section, the geometrical and aerodynamic properties are given as functions of the local radius, r, from the hub to the blade tip, $r = R$. Each blade section has the thickness dr as shown in **Figure 3.1(a)**. The axes x_1 and z define the plane of rotation.

The cross-sectional aerofoil element of the blade section at the local radius r is shown in **Figure 3.1(b)**. The cross-sectional element defines the blade element at radius r in (x_1, y_1)-plane. The local velocity, $V_{REL}(r)$, being relative to the rotating blade, is given by superimposing the axial velocity $V(1-a)$ and the rotational velocity $r\omega_M(1+a')$ at the rotor plane. The undisturbed wind speed is V and the tangential velocity of the rotating blade section is $r\omega_M$. The induced velocities in the rotor plane are $-aV$ and $+a'r\omega_M$, with the respective factors a and a'. These denote the velocities induced by the vortex system of the wind turbine.

The local angle of attack, α, is defined from the local flow angle, ϕ, the local twist of the blade, τ, and the global pitch angle, β, as (Sørensen and Kock, 1995):

$$\alpha = \phi - (\tau + \beta). \tag{3.1}$$

In the case of fixed-pitch wind turbines (often called stall-controlled), β is constant. In the case of pitch or active-stall controlled wind turbines, the pitch angle, β, is adjusted by the pitch control system. The local flow angle, ϕ, is the angle between the relative velocity, $V_{REL}(r)$, and the rotor plane:

$$\phi = \tan^{-1}\left(\frac{V \cdot (1-a)}{r \cdot \omega_M \cdot (1+a')} \right), \tag{3.2}$$

and the relative velocity itself is found by inspection of the velocity vectors shown in **Figure 3.1(b)**:

$$V_{REL} = \sqrt{\left(V \cdot (1-a)\right)^2 + \left(r \cdot \omega_M \cdot (1+a')\right)^2} \ . \tag{3.3}$$

The lift force, L, and the drag force, D, are computed with the use of the relations:

$$\begin{cases} L = \frac{1}{2} \cdot \rho_{AIR} \cdot V_{REL}^{'2} \cdot c \cdot C_L, \\ D = \frac{1}{2} \cdot \rho_{AIR} \cdot V_{REL}^{2} \cdot c \cdot C_D, \end{cases} \tag{3.4}$$

where ρ_{AIR} is the air density that is 1.225 kg/m^3 for standardised air, c is the local chord of the blade section and C_L and C_D are the lift and the drag coefficients, respectively. Note that the data for the local twist, τ, and the local chord, c, are given by the wind turbine manufacturer for each cross-sectional element of the blade. The coefficients C_L and C_D are given by the manufacturer as functions of the angle of attack, α, for each cross-sectional element of the blade.

The normal, F_N, and tangential, F_T, forces being parallel to the axes y_1 and x_1, respectively, and acting onto each blade section are found by inspection of **Figure 3.1(b)**.

$$\begin{cases} F_N = L \cdot \cos(\phi) + D \cdot \sin(\phi) , \\ F_T = L \cdot \sin(\phi) - D \cdot \cos(\phi) . \end{cases} \tag{3.5}$$

The thrust, T, the mechanical torque, T_M, the mechanical power, P_M, produced by the wind turbine rotor are computed according to:

$$\begin{cases} T = B \cdot \int_0^R F_N(r) \cdot dr , \\[2mm] T_M = B \cdot \int_0^R F_T(r) \cdot r \cdot dr , \\[2mm] P_M = \omega_M \cdot T_M . \end{cases} \tag{3.6}$$

The integration is performed from the blade root at $r = 0$ to the blade tip at $r = R$ where R denotes the rotor radius. To make the computation of the parameters T, T_M and P_M complete, it is however necessary to find the axial induction factor, a, and the rotational induction factor, a', for each cross-sectional blade element. This can be achieved with the use of the iterative routine described below.

The Prandtl's tip loss factor, F, is the correction factor that takes into account a finite number of the rotor blades, B. In the case of the three-bladed rotor $B = 3$. The Prandtl's factor is then defined as:

$$F = \frac{2}{\pi} \cdot \arccos\left(\exp\left(-\frac{B}{2} \cdot \frac{R-r}{r \cdot \sin(\phi)} \right) \right) . \tag{3.7}$$

The solidity of the given blade element is given by the expression:

$$\sigma = \frac{c \cdot B}{2\pi \cdot r}.$$

(3.8)

The normal and the tangential forces' coefficients C_N and C_T are defined analogously to the lift and the drag coefficients.

$$\begin{cases} C_N = \dfrac{F_N}{\frac{1}{2} \cdot \rho_{AIR} \cdot V_{REL}^2 \cdot c}, \\[2ex] C_T = \dfrac{F_T}{\frac{1}{2} \cdot \rho_{AIR} \cdot V_{REL}^2 \cdot c}. \end{cases}$$

(3.9)

When the axial induction factor, a, is below the critical induction factor a_C, this factor can be computed with the use of the expression:

$$a = \left(\frac{4F \cdot \sin^2(\phi)}{\sigma \cdot C_N} + 1 \right)^{-1}.$$

(3.10)

When the axial induction factor, a, exceeds the critical value a_C, the axial induction factor must be found from the equation:

$$4F \cdot \left(a_C^2 + (1 - 2a_C) \cdot a \right) = \frac{(1-a)^2 \cdot \sigma \cdot C_N}{\sin^2(\phi)}.$$

(3.11)

Commonly, the critical value a_C may be in the range from 0.2 to 0.33 (Hansen, 2000). The rotational induction factor, a', can be computed with the use of the expression:

$$a' = \left(\frac{4F \cdot \sin(\phi) \cdot \cos(\phi)}{\sigma \cdot C_T} - 1 \right).$$

(3.12)

The iterative routine to compute the induction factors a and a' is illustrated in **Figure 3.2**. Note that this routine produces the stationary values of the induction factors a and a', when the rotor is in equilibrium.

3.1.1 Power coefficient

Instead of the BEM method which requires information of the geometrical and aerodynamic data of the rotor blade profiles, the rotor mechanical power can be computed with the use of the power coefficient curves. Such curves are given by the wind turbine manufacturer as $C_P(\lambda,\beta)$ characteristics, where C_P denotes the power coefficient, λ is the tip-speed-ratio and β is the pitch angle. The tip-speed-ratio is defined as:

$$\lambda = \frac{\omega_M \cdot R}{V} .$$

(3.13)

Note that $C_P(\lambda,\beta)$ characteristics can also be computed with the use of the BEM method for any types of wind turbines (Sørensen and Kock, 1995; Younsi et al., 2001).

$$\begin{cases} C_P(\lambda,\beta) = \dfrac{P_M(V,\omega_M,\beta)}{P_V} , \\ P_V = \frac{1}{2} \cdot \rho_{AIR} \cdot V^3 \cdot \pi \cdot R^2 . \end{cases}$$

(3.14)

Here, the array of P_M is computed at fixed values of tip-speed-ratios, λ, and pitch angles, β. P_V is the wind power available in the swept rotor area of πR^2.

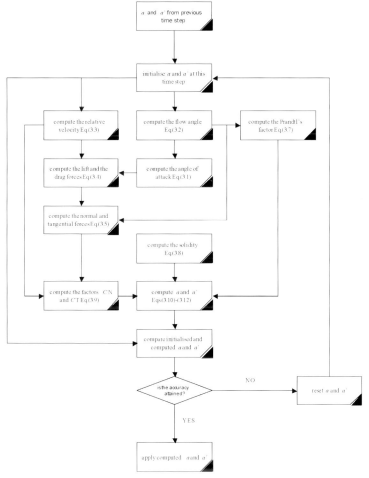

Figure 3.2   An iterative routine to compute the induction factors a and a'.

Note that the power coefficient of any wind turbine cannot exceed its theoretical Betz limit $C_P^{MAX} = 16/27 \approx 0.59$. This Betz limit is obtained for stationary operation of the wind turbines (Pretlove and Mayer, 1994; Walker and Jenkins, 1997; Hansen 2000). **Figure 3.3** shows the $C_P(\lambda,\beta)$ characteristics computed with the use of the BEM method for a given 2 MW wind turbine located in Denmark.

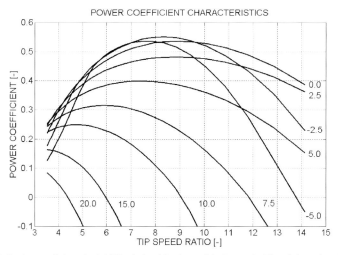

Figure 3.3 $C_P(\lambda,\beta)$ characteristics of a 2 MW wind turbine located in Denmark. The pitch angle values are shown at each characteristic.

When such characteristics are given, Eq.(3.13) and Eq.(3.14) allow the computation of the mechanical power, P_M, in any stationary operational point of the rotor, using the $C_P(\lambda,\beta)$ characteristics:

$$P_M = \tfrac{1}{2} \cdot \rho_{AIR} \cdot V^3 \cdot \pi \cdot R^2 \cdot C_P(\lambda,\beta) . \tag{3.15}$$

In this way, information about the blade geometry and its aerodynamic properties is not needed. With regard to wind turbine generator operation, the BEM method and the use of the $C_P(\lambda,\beta)$ characteristics will produce the same result for P_M applied to the generator shaft system. The use of the $C_P(\lambda,\beta)$ characteristics to produce the rotor power, P_M, is valid so long as the rotor is in equilibrium and unsteady inflow phenomena around the rotor do not occur. This is a reasonable assumption in many investigations of power system stability.

In the case of stall-controlled wind turbines, the pitch angle β is fixed. The rotor aerodynamics can be represented with the use of a single $C_P(\lambda)$ characteristic. The mechanical power of stall-controlled wind turbines is then given by the relation:

$$P_M = \tfrac{1}{2} \cdot \rho_{AIR} \cdot V^3 \cdot \pi R^2 \cdot C_P(\lambda) . \tag{3.16}$$

3.1.2 Fixed-speed operation

The rotor speed, ω_M, of wind turbines equipped with induction generators with a short-circuited rotor circuit is fixed to the electric system speed of the power grid in which the turbines feed. This implies that fixed-speed wind turbines, once started, always rotate at almost the same speed regardless of the wind speed, V. In normal power grid operations, the rotor speed may only vary according to the induction generator slip variations. The generator slip, s, at rated operation of the wind turbine may be in the range of 1 to 2% depending on induction generator parameters.

Such fixed-speed wind turbines are only able to operate at maximum efficiency at one particular wind speed. In the optimisation algorithm, the most probable wind speed in the installation area plays a part. The optimisation requires that the power coefficient, C_P, must reach a maximum at the wind speed above the most probable wind speed because the mechanical power, P_M, is a cubic function of the wind speed, V, see Eq. (3.16).

Note that fixed-speed wind turbines, installed in the same area and so in the same wind conditions, will have similar values of optimised tip-speed-ratios, λ_{OPT}, which are independent from the rated power, P_{RAT}. At optimised operation, Eq.(3.16) gives:

$$\begin{cases} P_M^{OPT} = \frac{1}{2} \cdot \rho_{AIR} \cdot V_{OPT}^3 \cdot \pi R^2 \cdot C_P^{OPT}, \\ \lambda_{OPT} = \dfrac{\omega_M \cdot R}{V_{OPT}}. \end{cases} \tag{3.17}$$

Since the optimised power coefficient, C_P^{OPT}, the optimised tip-speed-ratio, λ_{OPT}, and the optimised wind speed, V_{OPT}, are almost the same among different wind turbines located at the same location, Eq.(3.17) leads to two interesting conclusions.

1) To increase power output of fixed-speed wind turbines, the rotor radius, R, must be increased as P_M is proportional to R^2. When the rotor radius, R, has been increased, then the rotor speed, ω_M, can be reduced due to optimisation. **Table 3.1** illustrates this using data of fixed-speed wind turbines from the manufacturer Siemens.

2) To increase performance, fixed-speed wind turbines are often made to operate at two different fixed speeds. Such wind turbines are therefore called two-speed wind turbines and equipped with two induction generators, or with a single induction generator with changeable pole-pairs. **Table 3.2** gives an example of such a two-speed wind turbine available from the manufacturer Siemens.

Now consider that a 2 MW wind turbine located in Denmark is optimised to wind speed V_{OPT} =8 m/s. This implies that the mechanical power versus rotor speed curve $P_M(\omega_M)$ of this wind turbine has a maximum at 8 m/s. The $P_M(\omega_M)$ curves of this wind turbine are plotted in **Figure 3.4** and the maximum is reached at the optimised (fixed) rotor speed, which is 21.5 rev./min. for the given wind turbine.

At other wind speeds, the $P_M(\omega_M)$ curves for the fixed-speed wind turbine do not relate to the optimised operation. Further optimisation of the power output, P_M, can only be reached with the use of pitch angle control.

Rated power, kW	Rated rotor speed, rev./min.	Rotor radius, m
600	27	22
1300	19	31
2000	17	38

Table 3.1   Relation between the rotor radius and the rotor speed of fixed-speed wind turbines having different rated powers. Source: Siemens.

Specification	Generator data		Rated rotor speed, rev./min.
	Rated power, kW	Pole-pairs	
Siemens 1300 kW	1300	2	19
	250	3	13

Table 3.2   A wind turbine with two fixed speeds. Source: Siemens.

3.1.3   Variable-speed operation

In the case of variable-speed operations, wind turbines will produce more mechanical power, P_M, than in the case of fixed-speed operations (Zinger and Muljadi, 1997). Increased power output is reached by control of the rotor speed, ω_M, in accordance to the wind speed, V. The power gain is significant when the wind speed is below the optimised wind speed, V_{OPT}, of the fixed-speed operational mode. In this case, the variable-speed wind turbine is in sub-synchronous operation (Akhmatov, 2002(a)). The term "sub-synchronous operation" implies that the rotor speed, ω_M, is below the synchronous speed, i.e. 21.5 rev./min. in the example above, see **Figure 3.4**.

Figure 3.4 also includes a comparison between the fixed- and the variable-speed operation regimes for the same wind turbine rotor. However, in sub-synchronous operation, the wind turbine is controlled so that the power coefficient characteristics, C_P, reach the local maximum, C_P^{OPT}, at any wind speed. The other advantage of sub-synchronous operation is that the thrust, T, which represents the mechanical loads applied to the rotor construction, is reduced. The simulated curves of the $C_P(\lambda,\beta)$ characteristics and the thrust, T, are shown in **Figure 3.5** and **Figure 3.6**, respectively. In sub-synchronous operation, aerodynamic noise is also reduced compared to fixed-speed operation, since rotor speed is reduced.

The curves in **Figures 3.4** to **3.6** are computed at a fixed blade-angle, $\beta = 0$, and pitching in strong wind is not taken into account.

The power output, P_M, of variable-speed wind turbines can also be increased at wind speeds above the optimised wind speed, V_{OPT}, when compared to fixed-speed operation. This power increase is reached because optimisation of the power coefficient characteristics, C_P, continues in strong winds by running the rotor at super-synchronous speeds. The term "super-synchronous speed" implies that the rotor speed is larger than the synchronous rotor speed, e.g. larger than 21.5 rev./min. for the 2 MW wind turbine example. The optimised power output and the C_P characteristics at super-synchronous rotor speeds are illustrated in **Figures 3.4** and **3.5**, respectively.

The thrust, T, applied to the rotor construction and the tip-speed, $\omega_M R$, will increase at increasing wind speed, V, and increasing rotor speed, ω_M. For the thrust, this progressive increase at super-synchronous rotor speed is shown in **Figure 3.6**. At increasing rotor speed, the centrifugal forces applied to the generator windings and other rotating parts of the construction are also increasing. The wind turbine is a self-bearing construction, which must operate safely and without damage, therefore the thrust and the other forces loading the wind turbine construction must not be excessive.

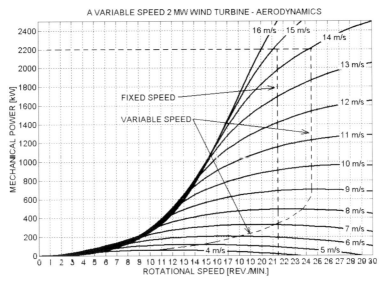

Figure 3.4 Mechanical power, P_M, versus rotor speed, ω_M, characteristics for a 2 MW wind turbine plotted at different wind speeds. The rotor diameter is 61 m and the curves were computed using the BEM-method. Reprinted from Ref. (Akhmatov, 2002(a)), Copyright (2002), with permission from Multi-Science Publishing Company.

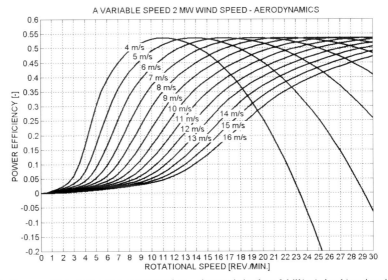

Figure 3.5 Power coefficient, C_P, versus rotor speed, ω_M, characteristics for a 2 MW wind turbine plotted at different wind speeds. The curves were computed using the BEM-method. Reprinted from Ref. (Akhmatov, 2002(a)), Copyright (2002), with permission from Multi-Science Publishing Company.

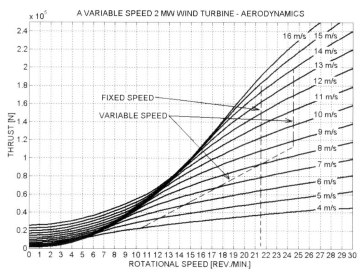

Figure 3.6 Thrust, T, versus rotational speed, ω_M, curves for a 2 MW wind turbine plotted at different wind speeds. The curves were computed using the BEM-method. Reprinted from Ref. (Akhmatov, 2002(a)), Copyright (2002), with permission from Multi-Science Publishing Company.

Increase of the tip-speed at increasing rotor speed will also act upon increase aerodynamic noise generated by the rotor. The tip-speed of modern wind turbines is limited to around 80 m/s. In other words, the rotor speed, ω_M, may not increase without limit, though the use of the upper rotor speed will imply that the mechanical power, P_M, is also limited.

Variable-speed wind turbines are, usually, equipped with pitch control. This pitch control is, in the first place, applied for optimisation of the power output, P_M, at a given wind speed. This optimisation is important in light wind, where the pitch control sets the pitch angle, β, in a way to increase the mechanical power, P_M. In strong wind, the pitch control is applied to limit the power output, P_M, to the rated power, P_{RAT}. The use of the pitch control in strong wind also reduces mechanical loading applied to the rotor construction.

Variable-speed operation makes it possible to increase the annual energy production by approximately 5% (Vestas, 2001).

3.1.3.1 Range of rotor speed

In variable-speed operation, the range of the rotor speed is chosen with regard to the following.

1) When the wind speed is below the optimised wind speed level, V_{OPT}, the rotor speed is in the sub-synchronous range to optimise the C_P characteristics and to maximise the mechanical power, P_M.

2) When the wind speed is slightly above the optimised wind, V_{OPT}, the rotor speed is in the super-synchronous range to optimise the C_P characteristics and to maximise the mechanical power, P_M.

3) In strong winds, the rotor speed, ω_M, is limited to keep the mechanical power at the rated value, reducing the mechanical loading on the rotor construction and the aerodynamic noise.

With the use of Eq.(3.13) for the tip-speed-ratio λ, the expression for the mechanical power of the wind turbine Eq.(3.16) can be rewritten according to (Neris et al., 1999; Song et al., 2000).

$$P_M = \frac{1}{2} \cdot \rho_{AIR} \cdot \pi R^5 \cdot \frac{C_P}{\lambda^3} \cdot \omega_M^3 . \tag{3.18}$$

In the following presentation, the optimisation algorithm is applied to reach the maximum power coefficient, C_P^{OPT}, at a given wind speed, V. In this algorithm, the rotor speed is replaced by the optimised rotor speed, ω_M^{OPT}, which relates to the given wind speed, V. The optimised tip-speed-ratio which corresponds to the maximum power coefficient, C_P^{OPT}, will then be given as:

$$\lambda_{OPT} = \frac{\omega_M^{OPT} \cdot R}{V} . \tag{3.19}$$

By insertion of Eq.(3.19) into Eq.(3.18), the maximum mechanical power is given by.

$$P_M^{OPT} = \frac{1}{2} \cdot \rho_{AIR} \cdot \pi R^5 \cdot \frac{C_P^{OPT}}{\lambda_{OPT}^3} \cdot \omega_M^3 = K_W \cdot \omega_M^3, \tag{3.20}$$

where K_W is a construction dependent coefficient.

Eq.(3.20) shows that the maximum mechanical power is a cubic function of the rotor speed, ω_M. According to Eq.(3.16), the mechanical power is also a cubic function of the wind speed, V. As can be seen, optimised rotor speed is proportional to the wind speed. This relation gives the optimised operational range of variable-speed wind turbines.

For a 2 MW wind turbine located in Denmark, the optimised rotor speed is 10.5 rev./min. at the wind speed of 4 m/s. Below this wind speed, the rotor speed is kept constant. When the wind speed has reached 9 m/s, the optimised rotor speed becomes 24.5 rev./min. At winds above this wind speed, the static rotor speed is kept as 24.5 rev./min. In the wind speed range from 4 m/s to 9 m/s, the optimised rotor speed is a linear function of the wind speed (Papathanassiou and Papadopoulos, 1999; Krüger and Andresen, 2001; Hansen et al., 2004(b)). **Figure 3.7** shows the dependence of the rotor speed on the wind speed for the given variable-speed wind turbine. The (static) range of variable-speed operation, the rotor speed, ω_M, may vary between -50% and $+10\%$ with regard to the synchronous speed. Therefore this is often called a fully variable-speed operation.

The most significant increase in power output, P_M, in variable-speed wind turbines is reached using variable-speed regimes (optimisation of the rotor speed, ω_M). Further optimisation of the power output with the use of pitch control may give up to 0.5% extra to the annual energy production.

As in the case of fixed-speed wind turbines, to increase the power output of variable-speed wind turbines, the rotor radius, R, must be increased since P_M is proportional to R^2. When the rotor radius, R, is increased, then the rotor speed, ω_M, can be reduced due to the optimisation. **Table 3.3** illustrates this design aspect using data of 2.x variable-speed wind turbines from the manufacturer GE Wind Energy.

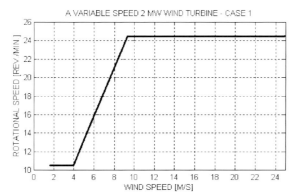

Figure 3.7   Rotor speed versus wind speed for a given variable-speed wind turbine. Reprinted from Ref. (Akhmatov, 2002(a)), Copyright (2002), with permission from Multi-Science Publishing Company.

Rated power, MW	Rotor speed range, rev./min.	Rotor radius, m
2.3	from 6.0 to 18.0	42
2.5	from 5.5 to 16.5	44
2.7	from 5.0 to 14.9	47

Table 3.3   Relation between the rotor radius and the rotor speed range of 2.x variable-speed wind turbines from GE Wind Energy. Source: GE Wind Energy.

3.1.4   Blade angle control

Blade-angle control is used for the following reasons in wind turbines.

1) Optimisation of wind turbines' power output, e.g. to produce maximum mechanical power at a given wind. The optimisation is only applied in light and moderate wind, when the wind speed, V, is below the rated wind, V_{RAT}.

2) Blade angle control must prevent the mechanical power exceeding the rated mechanical power in strong winds, when the wind speed is above the rated wind. This limits the mechanical power, P_M, and keeps it at the rated value in strong winds (Hansen, 2000; Akhmatov, 2002(a)).

3) The blade angle control is also applied to stop disconnected wind turbines from turning, thus operating as an aerodynamic break.

The above items relate mostly to wind turbines located in relatively small sites and connected to local distribution power grids. Large windfarms which are to be connected to transmission power grids must comply with the national Grid Codes of the system operators, for example (Eltra, 2000). According to this Grid Code (Eltra, 2000), large windfarms must be able to reduce their mechanical power output from any operational point to below 20% of their rated power in less than 2 s. This requirement can be met with the use of blade angle control, known as fast pitching (Akhmatov,

2001). Blade angle control can be used to stabilise the operation of large windfarms in the event of grid faults by preventing wind turbines over-speeding (Akhmatov, 2001). Blade angle control is therefore an important control system in investigations of power system stability. Therefore this control mechanism must be part of dynamic wind turbine models used to investigate power system stability and compliance with the national Grid Codes.

3.1.4.1 Power optimisation by pitch control

More optimisation of mechanical power, P_M, and C_P -characteristics can be reached by ensuring the global pitch angle, β, is at its optimised position. The optimised position of β is derived with regard to the wind speed, V, and, in the case of variable-speed wind turbines, with regard to the optimised rotor speed, ω_M^{OPT}. For variable-speed operation, the optimisation process using the BEM-method is illustrated in **Figure 3.8**. In this case, optimisation is performed so as to reach the maximum value of power coefficient, C_P. The power coefficient and then the mechanical power of variable-speed wind turbines can therefore be optimised by adjusting the pitch angle, which then gives a slight increase of the power gain. Note that the power output optimisation is viable only when the wind speed is below the rated wind speed, which in the example is 14 m/s for a 2 MW wind turbine located in Denmark.

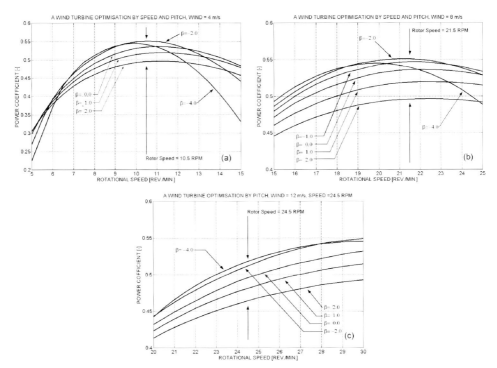

Figure 3.8   Optimisation of power output by blade angle control: **(a)** - wind speed of 4 m/s and rotor speed of 10.5 rev./min. (sub-synchronous), **(b)** - wind speed of 8 m/s and rotor speed of 21.5 rev./min. (synchronous), **(c)** - wind speed of 12 m/s and rotor speed of 24.5 rev./min. (super-synchronous).

When the rated wind speed is reached or exceeded, pitch control is used to keep the power at the rated value independent of the incoming wind speed. In this scenario, the pitch angle is turned into a position which results in reduction of the power coefficient, C_P. Hence, the wind turbine is not at its optimised operation condition when the wind speed is above the rated value.

3.1.4.2 Generic versus optimised blade pitch

Optimisation of the power output of wind turbines is an important issue. The more power generated, the larger profit will be reached at the same wind conditions in a given location. When a dynamic wind turbine model is applied in investigations of power system stability, refined optimisation of wind turbines can be of inferior interest. At such considerations, the generic pitch model can be used instead of the optimised pitch.

With regard to the application of the generic pitch model, the following considerations are useful to make before proceeding with the pitch control model (Akhmatov, 2002(a)).

1) Optimisation of the pitch angle leads only to a slight increase of mechanical power output compared to the power gain reached with the use of variable-speed. The use of pitch angle manipulation may increase power output by a couple percent in light and moderate winds only.
2) Optimised pitch values are all found to be close to $0°$ and, for a given 2 MW wind turbine, these are slightly negative values.
3) Although detailed optimisation has been made, it is plausible to consider that the factual operational point of a wind turbine deviates a little from the optimised operational point. This may occur, for example, because of dynamic wind changes, dust and dirt on the rotor blades, and oscillations and torsion of the rotor blades. Therefore a small deviation of the pitch angle from the exact optimised value may be a valid consideration when defining an operational point of the wind turbine.

Representation of optimisation by the pitch in light and moderate winds can be omitted in dynamic models of fixed- as well as variable-speed wind turbines applied in investigations of power system stability. In this case, the exact value of optimised pitch angle in light and moderate winds is simply replaced by the value $\beta = 0°$ at any wind speed below the rated value (Akhmatov, 2002(a); Hansen et al., 2004(b)).

3.1.4.3 Pitch and active-stall

With regard to limitations of the mechanical power to the rated power in strong winds, blade angle control of wind turbines can be obtained in two different ways.

1) Pitch-control where the mechanical power, P_M, is reduced when the global pitch angle, β, increases. This principle is mostly applied in variable-speed wind turbines. Fixed-speed wind turbines may occasionally also be equipped with pitch control.
2) Active-stall-control where the mechanical power, P_M, is reduced when the global pitch angle, β, decreases. This control principle is commonly applied in fixed-speed wind turbines.

In pitch-controlled wind turbines, the angle of attack, α, decreases when the global pitch angle increases. As can be seen in **Figure 3.1(b)**, the blade nose is moved against the incoming wind, V, at increasing pitch, β. Then, the lift force, L, also decreases. This mechanism explains the reduction of mechanical power generated by the wind turbine at increasing pitch angle.

In the case of active-stall control, the angle of attack, α, increases with decreasing pitch, β. The blade section is moved across the incoming wind, V, as seen from **Figure 3.1(b)**. When the angle of attack becomes large, stall occurs. This reduces the lift force, L, and increases the drag force, D, which leads to a reduction of the mechanical power, P_M, produced by the rotor.

3.1.4.4 Control system model

The blade angle control model contains the initialisation routine as well as dynamic part. Under initialisation, the initial value of the blade angle, β, and its reference, β_{REF}, are found. As can be seen from the general description of blade angle control, the initial blade angle in light and moderate winds is the optimised pitch angle, β_{OPT}. In terms of generic representation, $\beta_{OPT} = 0°$ when the wind speed, V, is below the rated value, V_{RAT}.

The initial blade angle, β, and its reference, β_{REF}, at the rated operational point, when the wind speed exceeds the rated value, can be found with the use of the BEM method. In this presentation, the BEM method is modified to maintain the rated mechanical power in strong winds by adjusting the blade angle, β. $\beta(V)$ curves representing the initial blade angle versus the wind speed are given in **Figure 3.9** for the pitch and active-stall control regions of the same 2 MW wind turbine. When the wind speed, V, is below the rated value, V_{RAT}, the initial $\beta(V)$ curve relate to the optimised blade angle, β_{OPT}. The initial values of β in strong winds are, obviously, not optimised values.

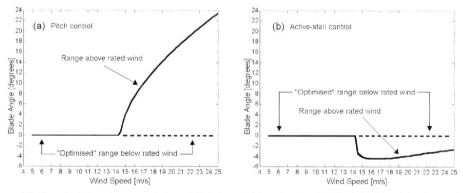

Figure 3.9 Generic pitch angle versus wind speed $\beta(V)$ characteristics of a given 2 MW wind turbine: **(a)** - when pitch-controlled, **(b)** - when active-stall controlled. Computed using the BEM method.

The $\beta(V)$ characteristic can be requested from wind turbine manufacturers together with other information about the rotor aerodynamics and control systems. The wind turbine manufacturer will normally deliver the fully optimised values of the pitch angle at a wind speed below the rated wind speed. In this case, the original values can be used in simulations or these can be replaced by zeros to perform simulations in a generic way.

The dynamic blade angle control can be organised in a generic way and, in the case of the pitch control, is given in **Figure 3.10**. In this control system, the blade angle, β, is controlled by the value X, which can be:

1) An electrical value. For instance, the active power, P_E, as described in (Hinrichsen, 1984; Akhmatov, 2001).
2) A mechanical value. For instance, the generator rotor speed, ω_G, as described in (Akhmatov, 2002(a); Akhmatov et al., 2003(a)).
3) A combination of electrical and mechanical values, according to (Akhmatov et al., 2003(a)).

The controlling value, X, is compared to its reference, X_{REF}. The error signal, X_{ERR}, is sent to the proportional-differential (PD) controller (optional, but gives better sensitivity) and then to the proportional-integral (PI) controller producing the reference value of the pitch angle, β_{REF}. This describes the generic, regular control system of the blade angle control. This control system is applied at normal wind turbine and the power grid operations (Akhmatov, 2001).

The reference value, β_{REF}, is kept in a range which depends on the control mode, between the optimised value, β_{OPT}, and the maximum pitch value, $\beta_{MAX} = 90°$, in the case of pitch-controlled wind turbines. However, it will be between the minimum pitch value, β_{MIN}, and the optimised value, β_{OPT}, in the case of active-stall controlled wind turbines.

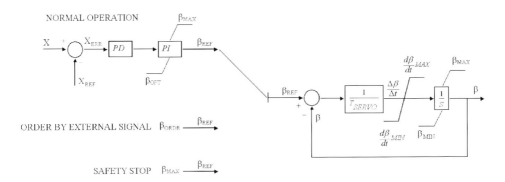

Figure 3.10 Generic blade angle control system in the case of pitch control. Active-stall control will be achieved when changing the upper limit of the PI-controller to β_{OPT}, its lower limit to β_{MIN}, and the reference value of the safety stop to β_{MIN}. Reprinted from Ref. (Akhmatov, 2003(b)), Copyright (2003), with permission from the copyright holder.

The restrictions applied to the reference value, β_{REF}, for the pitch and active-stall control, must be kept in mind.

1) In the case of pitch-controlled wind turbines, the (dynamic) reference value, β_{REF}, must not be below the optimised value, β_{OPT}, otherwise, there is a risk that the control system will go into the active-stall control domain.

2) In the case of active-stall control, the (dynamic) reference value, β_{REF}, must not increase beyond the optimised value, β_{OPT}, otherwise, there is a risk that the regular control system will go into the pitch control domain.

For representation of large windfarms, which must comply with the Danish Grid Code (Eltra, 2000), the regular blade angle control is extended with a control loop featuring the power ramp. The power ramp implies in this context the power reduction of the windfarm by an order given from the external system to the large windfarm. When the order to reduce the power output is given, the regular control system is disabled. The reference pitch angle is then set to the value $\beta_{REF} = \beta_{ORDR}$, which is the pitch angle corresponding to the desired power output. This desired power can, for example, be 20% of the rated power, inline with the Danish Grid Code (Eltra, 2000). The value of β_{ORDR} can also be computed knowing the wind speed and the rotor speed (Akhmatov et al., 2003(a)). In practice, values of β_{ORDR} can be reached from a look-up table generated by the manufacturer for each wind speed, $\beta_{ORDR}(V)$.

The large windfarm will continue operating with reduced power output so long as the order is given. When the order is cancelled, the regular control system is reset and restarted (Akhmatov et al., 2003(a)). The large windfarm then starts to regulate the power output according to the incoming wind resource.

Additionally, the safety stop control with the use of an aerodynamic break can be implemented (Snel and Schepers, editors, 1995). When safety stop is ordered, the regular control system is disabled. The reference pitch angle is set to $\beta_{REF} = \beta_{STOP}$ where $\beta_{STOP} = \beta_{MAX}$ in the case of pitch control, and $\beta_{STOP} = \beta_{MIN}$ in the case of active-stall control. When given, the order to stop cannot be cancelled.

The pitch servo system compares the measured pitch angle, β, to the reference pitch angle, β_{REF}, and corrects the error signal. Usually a first-order servo model is sufficient in investigations of power system stability (Akhmatov, 2002(a); Hansen et al., 2004(b)). However, the use of a more detailed pitch servo model can be required. Ref. (Hansen and Bindner, 1999) describes a second-order servo model used in investigations of responses of a given wind turbine.

For getting a realistic response of the generic, regular pitch control, a number of delay mechanisms must be implemented in control system models (Akhmatov, 2002(a)). Such delays represent sampling and filters dampening natural frequencies in the wind turbine construction (Leith and Leithead, 1997; Akhmatov, 2002(a)). Such delays can take seconds which makes the regular pitch control a slow-acting system when compared to the characteristic time of transient events in the power grid.

Note that models of blade angle control may additionally include blocks representing compensation of rotor non-linearity (Leith and Leithead, 1997). Such representations are component-specific and complex. In the generic model of the pitch control to be applied in investigations of power system stability, such complex representations may be omitted. The exception is when a component-specific model is required (as part of the model ordered by the wind turbine manufacturer for specific studies).

3.1.5 Dynamic inflow phenomena

Until now, it has been considered that the rotor aerodynamics could be described with the use of the BEM method or with the use of the $C_P(\lambda,\beta)$ characteristics. These methods are however static and based on the assumption that the rotor and the wind profile around the rotor are in equilibrium. The assumption of reaching equilibrium of the rotor operation is the same as to say that the wind profile formed around the rotor is able to change immediately at changes of the pitch angle, β. In other words, it is assumed that the wind profile around the rotor would transit from one present state of equilibrium stepwise to another state of equilibrium without delay (caused by a transition process). This assumption will however be in disagreement with the physics of the rotor (Øye, 1986; Snel and Shepers, 1992; Hansen et al. 2005).

The transition process between two states of equilibrium is continuous (with a characteristic time constant), but not immediate (i.e. stepwise). The transition process is defined by dynamic inflow phenomena (Øye, 1986). Dynamic inflow phenomena manifest themselves by overshoots occurring in the mechanical torque of the rotor during pitching. This phenomenon was first observed by Øye (1986) on pitch-controlled, medium-sized wind turbines in Denmark. The same behaviour was seen in measurements of a 2 MW pitch-controlled wind turbine equipped with an induction generator (Snel and Schepers, 1992).

In Ref. (Snel and Schepers, 1992), the 2 MW wind turbine is subject to a "step" of pitch angle, achieved by a sudden change in the pitch reference, β_{REF}. The pitch angle is then adjusted by the pitch servo with the servo time constant $T_{SERVO} = 0.25$ s and at a pitch rate limit of 7.2 °/s. The measured curve of the mechanical torque corresponding to the study (Snel and Schepers, 1992) is shown in **Figure 3.11**. This study case corresponds to an operational point at a wind speed of 8.7 m/s and initial pitch angle of $0°$. At time $t = 1$ s, the pitch reference was set to $+3.7°$, and at time t $= 30$ s, the pitch reference was again set to $0°$. As can be seen, the measured torque contains notable overshoots during pitching. Such overshoots cannot be explained in terms of the static BEM-method. To compute the overshoots, the BEM-method, which is a static method, must be revised. Ref. (Snel and Schepers, editors, 1995), for example, describes several methods to compute such a transition process in the wind profile around the rotor.

Here, the engineering model suggested by Øye (1986) is used as this model requires only a slight revision of the BEM method. In the Øye model, an introduction of the characteristic time constants of the transition process is required (Øye, 1986) which are in the order of $2R/V$ and $c/(\omega_M R)$. In the Øye model, the characteristic time constants appear in lags of induced velocities (the induction factors a and a' at a stationary wind speed and an almost stationary rotor speed). According to the computation diagram shown in **Figure 3.2**, the lags with the time constants $2r/V$ and $c/(\omega_M r)$ must appear between the induction factors a and a' (from the previous time step) and the factors a and a' computed by the iterative routine at the end of the process.

The model of the wind turbine with representation of the dynamic inflow phenomena, and using the Øye model, is implemented in the simulation tool PSS/E^{TM} as a user-written model (Akhmatov, 2003(c)). The simulated mechanical torque of the wind turbine, corresponding to the study case of Ref. (Snel and Schepers, 1992), is shown in **Figure 3.12**. The measured and simulated curves are in good agreement.

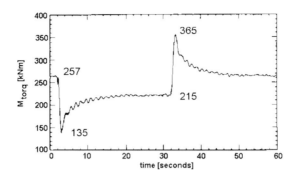

Figure 3.11 Measured mechanical torque of the wind turbine at pitch change. Reprinted from Ref. (Snel and Schepers, 1992), Copyright (1992), with permission from Elsevier.

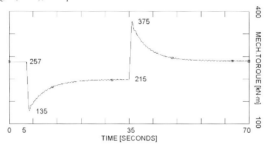

Figure 3.12 Computed mechanical torque, T_M, of the wind turbine as a response to a pitch change. Reprinted from Ref. (Akhmatov, 2003(c)), Copyright (2003), with permission from Multi-Science Publishing Company.

Note that the oscillations in the measured torque of the wind turbine, shown in **Figure 3.11**, are caused by the oscillations in the wind turbine construction. Similar oscillations are seen in the simulated behaviour in **Figure 3.12**. This oscillating behaviour is present in the simulated curve because it is computed using the Øye model combined with the elements of the aeroelastic code of the rotor as described in Ref. (Akhmatov, 2003(c)). The aeroelastic code is beyond the scope of this book and will not be discussed here.

3.1.5.1 Influence of pitch and active-stall

The validated rotor model will now be applied to evaluate where such dynamic inflow phenomena may play a part and where the $C_P(\lambda,\beta)$ characteristics can be applied without loss of accuracy. It is necessary to clarify whether the wind turbine is with pitch or active-stall control. The influence of the control mode on the response of the rotor with regard to the dynamic inflow phenomena is shown in **Figure 3.13**. The plotted curves in each case are modelled using the model with representation of dynamic inflow phenomena, the Øye model, and with the use of the $C_P(\lambda,\beta)$ characteristics which are derived assuming equilibrium. Note that the model with the representation of dynamic inflow phenomena requires more complex computations and time than the model using the $C_P(\lambda,\beta)$ characteristics. Therefore it is necessary to clarify (i) where the model with dynamic inflow phenomena must be applied to reach sufficient accuracy in the computed results and (ii) where the $C_P(\lambda,\beta)$ characteristics can be applied without loss of accuracy.

The simulation cases were chosen with regard to the Danish Grid Code and its requirement that the wind turbine can be ordered to the power ramp. In the simulation cases, the given 2 MW wind turbine located in Denmark is initially at its rated operation and ordered to reduce its power to 20% of its rated power. After a while, the order is cancelled and the wind turbine returns to the rated operational point.

As can be seen, there is only a marginal difference between the rotor responses modelled with the representation of dynamic inflow phenomena and with the use of the $C_P(\lambda,\beta)$ characteristics when the active-stall control is applied. This result is reached because the rotor blades are pitched across the incoming wind when the active-stall control is applied. Then, the wind profile around the rotor does not change significantly which only gives small overshoots in the torque.

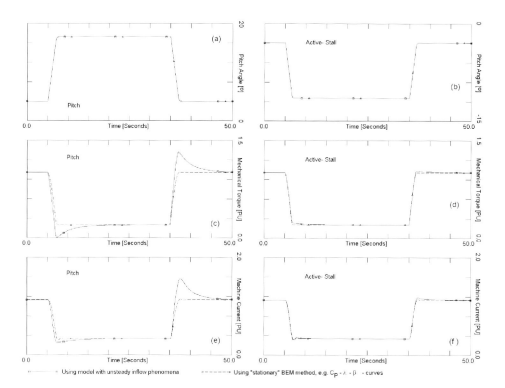

Figure 3.13   The response of a 2 MW wind turbine equipped with an induction generator computed using the model with dynamic inflow phenomena and $C_P(\lambda,\beta)$ characteristics. The response is reached with the use of pitch and active-stall control: **(a)** and **(b)** – pitch angle in the case of pitch control with a maximum pitch rate of 8 °/s, and active-stall control with a maximum pitch rate of 5 °/s, **(c)** and **(d)** – mechanical torque of the rotor, **(e)** and **(f)** – the generator current. Reprinted from Ref. (Akhmatov, 2003(b)), Copyright (2003), with permission from the copyright holder.

When pitch control is applied, a significant discrepancy is seen due to the notable overshoots in the mechanical torque, T_M. Such overshoots are caused by dynamic inflow phenomena. When pitch control is applied, the rotor blades are pitched away from incoming wind. This may lead to a significant change in the wind profile around the rotor and then dynamic inflow phenomena may occur.

3.1.5.2   Relevance of dynamic inflow

With regard to the accuracy of rotor modelling, the rotor of fixed-speed, stall and active-stall controlled wind turbines can be modelled using the $C_P(\lambda)$ and $C_P(\lambda,\beta)$ characteristics, respectively, without loss of accuracy. When the active-stall control is applied, rotor blade pitching does not result in any significant overshoot of the rotor torque, T_M. The rotors of pitch-controlled wind turbines can also be modelled using the $C_P(\lambda,\beta)$ characteristics when accuracy of the mechanical torque computation is not critical. This simplified representation may however introduce restrictions on the model application, such as the restriction on the pitch rate.

When accuracy of the computed rotor torque is required, it is preferred to model the rotors of pitch-controlled wind turbines with representation of dynamic inflow phenomena. In this way, overshoots in mechanical torque at (relatively fast) pitching will be present. The simplified model using the $C_P(\lambda,\beta)$ characteristics represents the rotor in equilibrium and does not produce such overshoots. Note that the mechanical torque, T_M, of the pitch controlled wind turbine shown in **Figure 3.13** is computed using a wind speed of 15 m/s and rotor diameter of 61 m. The characteristic time constants of the transition process influence the magnitude of the overshoots in the rotor torque. As expected, such overshoots will be ever larger when the wind speed is lower and the rotor diameter is larger. Rotor diameters of pitch-controlled wind turbines can be larger than 100 m, which may increase the magnitude and duration of such overshoots in the mechanical torque compared to the behaviour shown in **Figure 3.13**.

Only at sufficiently small pitch rates, overshoots in mechanical torque, T_M, are almost eliminated. **Figure 3.14** compares the mechanical torque of a 2 MW wind turbine at different pitch rates. When the pitch rate is 2 °/s, the overshoot in mechanical torque is reduced significantly. In this case, the rotor of the pitch-controlled wind turbine can be modelled using the $C_P(\lambda,\beta)$ characteristics without loss of accuracy.

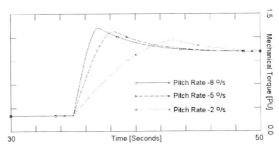

Figure 3.14. Mechanical torque of a pitch-controlled wind turbine when the power increases from 20% to 100% of the rated power at different pitch rate limits. Reprinted from Ref. (Akhmatov, 2003(b)), Copyright (2003), with permission from the copyright holder.

Dynamic inflow phenomena are not directly relevant for investigations into power system stability. However, an understanding of such mechanisms is useful when modelling wind turbines at critical events with the use of simplified models. It must be kept in mind that dynamic inflow phenomena will introduce restrictions on rotor operations in pitch controlled wind turbines. For example, the pitch rate of pitch controlled wind turbines will be limited to reduce the risk of mechanical torque overshoots in rotor.

Although active-stall controlled wind turbines do not experience such overshoots, the pitch rates of these wind turbines will be limited due to sensitivity of the mechanical torque to the pitch angle. This can be seen from the $\beta(V)$ characteristic shown in **Figure 3.9(a)**, for example.

3.1.6   Wind versus power characteristics

For a given wind turbine, the wind versus power characteristics show a relation between the wind speed and the mechanical power, $P_M(V)$, or between the wind speed and the electrical power, $P_E(V)$. The $P_E(V)$ characteristics computed or measured at standardised conditions are often given by wind turbine manufacturers. The $P_E(V)$ characteristic is of interest to the customer because it shows the value of electrical power which the wind turbine supplies to the grid at given wind speeds.

Mechanical $P_M(V)$ characteristics can be computed using the BEM method or the $C_P(\lambda,\beta)$ characteristics together with the optimised rotor speed versus wind speed characteristic, $\omega_M^{OPT}(V)$, in the case of the variable-speed wind turbines, and the pitch versus wind speed characteristic, $\beta(V)$, in the case of variable-pitch wind turbines. The computation of the $P_M(V)$ characteristics using the BEM method is given by Eq.(3.1) and Eq.(3.12) whereas the induction factors a and a' are computed according to the routine shown in **Figure 3.2**.

The computational routine using the $C_P(\lambda,\beta)$ characteristics is shown in **Figure 3.15**. The $P_M(V)$ characteristic of a fixed-speed, variable-pitch 2 MW wind turbine is shown in **Figure 3.16.** The characteristic starts with the cut-in wind speed, V_{IN}, (5 m/s for the given wind turbine). The cut-in wind speed is the wind speed at which the wind turbine is programmed to start operating. When the wind speed is below V_{IN}, there is very little energy in the wind and operation of the wind turbine is unviable or impossible. The rated wind speed is the minimum wind speed at which the wind turbine produces the rated electrical power. The rated electrical power, P_{RAT}, which is 2,0 MW for the example wind turbine, is reached at the rated wind speed, V_{RAT}, 14.4 m/s. The characteristic ends at the cut-out wind speed, V_{OUT}, (25 m/s for most of wind turbines). The cut-out wind speed is the wind speed at which the wind turbine shuts down due to safety reasons. Though such high-speed winds are energetic, they are experienced so rarely at most of locations that the energy wasted is small during periods when the wind turbine shuts down.

Note that the "mechanical" $P_M(V)$ characteristics is shown above the "electrical" $P_E(V)$ characteristics due to the presence of the mechanical and the electrical losses in a physical wind turbine. Such losses are due to friction, and resistive losses in the generator. The total losses, ΔP, can be estimated using a polynomial expression with regard to the mechanical power.

$$\Delta P = a_P + b_P \cdot P_M + c_P \cdot P_M^2 .$$

$$(3.21)$$

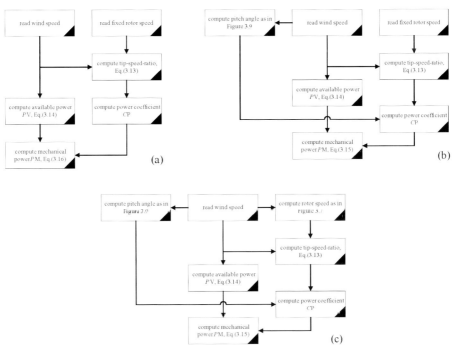

Figure 3.15 Schematic diagrams of the computation of wind versus power characteristic of a wind turbine: **(a)** - fixed-speed, stall controlled, **(b)** - fixed-speed with blade angle control, **(c)** - variable-speed, pitch controlled.

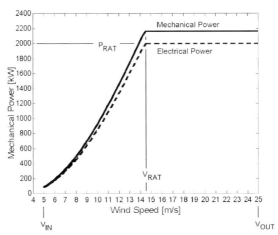

Figure 3.16 Wind versus power characteristics of a given 2 MW, fixed-speed, active-stall controlled wind turbine.

The coefficients in Eq.(3.21) can be estimated experimentally. For the given, 2 MW wind turbine, the coefficients are a_P =40 kW, b_P =1.852·10^{-2} and c_P =1.715·10^{-5} kW^{-1}. The $P_E(V)$ characteristic can be computed from the $P_M(V)$ characteristics with the use of Eq.(3.21).

$$P_E(V) = P_M(V) - \Delta P. \tag{3.22}$$

The $P_E(V)$ characteristic plotted in **Figure 3.16** was computed using Eq.(3.21) and Eq.(3.22).

3.1.7 Approximate rotor modelling

In the Western part of Denmark, there are about 4,500 wind turbines. Among those, there are a significant number of small and medium -sized, fixed-speed, stall controlled wind turbines. The rated power of those wind turbines is below 600 kW and often there are no available aerodynamic data such as the $C_P(\lambda)$ characteristics, although it is known about the cut-in wind speed, V_{IN}, the rated wind speed, V_{RAT}, the cut-out wind speed, V_{OUT} and the rated electrical power, P_{RAT}. This information can be sufficient to compute an approximate wind versus power characteristic, $P_E(V)$, and then an approximate $C_P(\lambda)$ characteristic (with respect to the electrical power) of the given wind turbine. Consider now that the wind versus power characteristic of a stall controlled wind turbine has a minimum at the cut-in wind speed, V_{IN}, a maximum at a wind speed slightly above the rated wind, which is denoted as V_{MAX}, and a minimum at the cut-out wind speed, V_{OUT}. From this consideration, the derivative of the $P_E(V)$ characteristic must be a third-order polynomial expression.

$$\frac{dP_E}{dV} = (V - V_{IN}) \cdot (V - V_{MAX}) \cdot (V - V_{OUT}) . \tag{3.23}$$

Introducing a new variable

$$u = V - V_{IN} , \tag{3.24}$$

this represents the coordinate shift from V to u (the cut-in wind speed in the u-axis is at 0 m/s). The solution of Eq.(3.23) with regard to the variable u is given by:

$$P_E(u) = A \cdot u + B \cdot u^2 + C \cdot u^3 + D \cdot u^4 . \tag{3.25}$$

The target is to define the coefficients A, B, C and D in terms of known parameters. These four coefficients can be found from the four conditions given below.

$$\begin{cases} \left(\dfrac{dP_E}{du}\right)_{u=0} = \left(A + 2B \cdot u + 3C \cdot u^2 + 4D \cdot u^3\right)_{u=0} = 0, \\[2mm] \left(\dfrac{dP_E}{du}\right)_{u=u_{MAX}} = \left(A + 2B \cdot u + 3C \cdot u^2 + 4D \cdot u^3\right)_{u=u_{MAX}} = 0, \\[2mm] \left(\dfrac{dP_E}{du}\right)_{u=u_{OUT}} = \left(A + 2B \cdot u + 3C \cdot u^2 + 4D \cdot u^3\right)_{u=u_{OUT}} = 0, \\[2mm] \left(P_E\right)_{u=u_{RAT}} = \left(A \cdot u + B \cdot u^2 + C \cdot u^3 + D \cdot u^4\right)_{u=u_{RAT}} = P_{RAT} \end{cases} \tag{3.26}$$

The coefficients are found as the solution of Eq.(3.26):

$$
\begin{cases}
A = 0, \\
B = 6 \cdot P_{RAT} \cdot \dfrac{u_{OUT} \cdot u_{MAX}}{E}, \\
C = -4 \cdot P_{RAT} \cdot \dfrac{u_{OUT} + u_{MAX}}{E}, \\
D = 3 \cdot P_{RAT} \cdot \dfrac{1}{E},
\end{cases}
\qquad (3.27)
$$

where the denominator, E, is expressed by:

$$
E = u_{RAT}^2 \cdot \left(3 \cdot u_{RAT}^2 - 4 \cdot u_{RAT} \cdot u_{MAX} - 4 \cdot u_{RAT} \cdot u_{OUT} + 6 \cdot u_{OUT} \cdot u_{MAX} \right). \qquad (3.28)
$$

Finally, the approximate expression of the $P_E(V)$ characteristic becomes:

$$
P_E(V) = P_{RAT} \cdot \frac{6 \cdot u_{OUT} \cdot u_{MAX} \cdot (V - V_{IN})^2 - 4 \cdot (u_{OUT} + u_{MAX}) \cdot (V - V_{IN})^3 + 3 \cdot (V - V_{IN})^4}{u_{RAT}^2 \cdot \left(3 \cdot u_{RAT}^2 - 4 \cdot u_{RAT} \cdot u_{MAX} - 4 \cdot u_{RAT} \cdot u_{OUT} + 6 \cdot u_{OUT} \cdot u_{MAX} \right)}, \quad (3.29)
$$

which is valid in the range of the cut-in wind speed, V_{IN}, to the cut-out wind speed, V_{OUT}. Beyond this range, $P_E(V)$ is zero. The other expressions in Eq.(3.29) are:

$$
\begin{cases}
V_{MAX} \approx \dfrac{2V_{RAT} + V_{OUT}}{3}, \\
u_{RAT} = V_{RAT} - V_{IN}, \\
u_{MAX} = V_{MAX} - V_{IN}, \\
u_{OUT} = V_{OUT} - V_{IN}.
\end{cases}
\qquad (3.30)
$$

The approximate expression of the $P_E(V)$ characteristic is compared to the measured characteristic of the Vestas V39, 500 kW wind turbine, as shown in **Figure 3.17**. The approximate expression Eq.(3.29) is found to be in good agreement with the measured characteristic of the 500 kW wind turbine.

The tip-speed-ratio, λ, can be computed at different values of wind speeds, V, using Eq.(3.13) and knowing the rated (fixed) rotor speed, ω_M, and the rotor radius, R. The C_P values at each wind speed, V, are computed according to:

$$
\begin{cases}
C_P(V) = \dfrac{P_E(V)}{P_V}, \\
\lambda = \omega_M \cdot R / V
\end{cases}
\qquad (3.31)
$$

where P_V is the power available in the rotor swept area due to incoming wind, V, and given by Eq.(3.14). When replacing the wind speed, V, by the related value of the tip-speed-ratio, λ, the approximate $C_P(\lambda)$ characteristic is gained with only the use of basic wind turbine data, such as the cut-in wind speed, rated wind speed, cut-out wind speed, rated electrical power, rated rotor speed and rotor radius. All such data are normally given or published by wind turbine manufacturers.

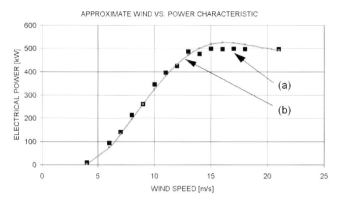

Figure 3.17   Validation of the approximate expression of the wind versus power characteristic: **(a)** - measured electrical power of a V39 500 kW wind turbine. Courtesy of Vestas Wind Systems, **(b)** - approximate characteristic computed at V_{IN} =3.5 m/s, V_{RAT} =14 m/s, V_{OUT} =21 m/s and P_{RAT} = 500 kW.

Table 3.4 shows the computed results of the approximate $C_P(\lambda)$ characteristic computed with the use of Eq.(3.29). Note that the $C_P(\lambda)$ characteristic in **Table 3.4** was computed with regard to the electrical power, P_E, i.e. the mechanical power minus mechanical and electrical losses in the wind turbine. The $C_P(\lambda,\beta)$ characteristics received from wind turbine manufactures can be measured or computed characteristics obtained with regard to the electrical power. It can therefore be useful to clarify with the manufacturer the origin of the $C_P(\lambda,\beta)$ characteristics received.

Wind, m/s	P_V, kW, Eq.(3.14)	P_E, kW, Eq.(3.28)	λ, Eq.(3.30)	C_P, Eq.(3.30)
4	46.83	3.61	13.78	0.08
6	158.04	74.53	9.19	0.47
7	250.97	131.87	7.88	0.53
8	374.62	195.88	6.89	0.52
9	533.40	261.62	6.13	0.49
10	731.69	324.96	5.51	0.44
11	973.88	382.57	5.01	0.39
12	1,264.35	431.96	4.59	0.34
13	1,607.52	471.40	4.24	0.29
14	2,007.75	500.00	3.94	0.25
15	2,469.44	517.68	3.68	0.21
16	2,996.99	525.14	3.45	0.18
17	3,594.78	523.93	3.24	0.15
18	4,267.20	516.36	3.06	0.12
21	6,776.15	491.02	2.63	0.07

Table 3.4   Computed approximate $C_P(\lambda)$ characteristic for a stall controlled, fixed-speed 500 kW wind turbine.

3.1.8   Initialisation and application of the rotor model

Before dynamic computations can be started, the rotor model must be initialised. This initialisation implies that the mechanical power (or the mechanical torque) corresponding to the rotor operation in steady-state must be found as part of a more complex wind turbine model. The mechanical power of the rotor must comply with operational conditions of other wind turbine components which are represented by their respective models. The mechanical power of the rotor can be initialised either from the wind speed, V, if it is given, or from the active power of the induction generator, P_E, which is known from the load-flow solution of the power grid. There are several mechanisms of power losses between the generator terminals with the active power P_E and the rotor with the mechanical power P_M. Such losses are not always possible to compute exactly. Therefore there can be an initial mismatch between the mechanical power computed from the rotor aerodynamics, P_{MV}, and the mechanical power computed from the generator side of the wind turbine model, P_{ME}, with inclusion of such losses. Schematically this initial mismatch is given as:

$$\left(Aerodynamics\right) \rightarrow P_{MV}(t=0) \neq P_{ME}(t=0) = P_E(t=0) + \Delta P(t=0) \leftarrow \left(Load - flow\right) \quad (3.32)$$

When the dynamic wind turbine model is applied to investigations of power system stability, the mechanical power of the rotor must comply with operational conditions at the generator terminals. Therefore, the initial mechanical power of the rotor is factually determined from the active power of the wind turbine generator, P_E.

$$P_M(t=0) = P_{ME}(t=0) = P_E(t=0) + \Delta P(t). \quad (3.33)$$

In this case, some inaccuracy in the determination of the initial wind speed, $V(t=0)$, and the mechanical power initialised from the wind speed, e.g. aerodynamics, $P_{MV}(t=0)$ is accepted so long as this inaccuracy is within a relatively small range of ε.

$$\left| \frac{P_{MV}(t=0) - P_{ME}(t=0)}{P_{MV}(t=0) + P_{ME}(t=0)} \right| \leq \varepsilon. \quad (3.34)$$

In Eq.(3.34) the mechanical powers $P_{MV}(t=0)$ and $P_{ME}(t=0)$ account for a single wind turbine. When the rotor model is initialised within sufficient accuracy ε according to Eq.(3.34), the dynamic computations of the mechanical power can be performed as:

$$P_M(t) = P_{ME}(t=0) \cdot \frac{P_{MV}(t)}{P_{MV}(t=0)}, \quad (3.35)$$

where $P_{MV}(t)$ is computed at any time t with the use of the aerodynamic rotor model for a single wind turbine. This computation can be made using the BEM method, the $C_p(\lambda,\beta)$ characteristics or the model with representation of the dynamic inflow phenomena. The initial power $P_{ME}(t=0)$ and the dynamic mechanical power $P_M(t)$ are computed for a single wind turbine or for a group of wind

$$\begin{cases} 2H_M \cdot \dfrac{d\omega_M}{dt} = T_M - T_G - D_M \cdot \omega_M \ , \\[2mm] 2H_G \cdot \dfrac{d\omega_G}{dt} = T_G - T_E - D_G \cdot \omega_G \ , \\[2mm] \dfrac{d\theta_S}{dt} = \omega_0 \cdot (\omega_M - \omega_G) \ , \end{cases} \qquad (3.46)$$

where D_M, D_G and D_S are damping coefficients of the rotor, generator rotor and shaft, respectively, and the mechanical torque of the rotor, T_M, and the mechanical torque applied to the generator rotor shaft, T_G, are given as:

$$\begin{cases} T_M = \dfrac{P_M}{\omega_M}, \\[2mm] T_G = K_S \cdot \theta_S - D_S \cdot (\omega_G - \omega_M) \end{cases} \qquad (3.47)$$

The mechanical power of the rotor, P_M, is computed using Eq.(3.35) or Eq.(3.36) of the rotor aerodynamics applying a suitable per unit system. The electrical torque, T_E, is the result of the generator model. The states of the shaft system model are initialised from Eq.(3.46) when the derivatives of the states are set to zero.

In the case of fixed-speed wind turbines equipped with conventional induction generators, the initial generator rotor speed, $\omega_G(t=0)$, is found from the generator model initialisation as shown in **Section 4.2.6.1**. The initial rotor speed in p.u., $\omega_M(t=0)$, is set equal to the initial generator speed in p.u. (minus slip).

In the case of variable-speed wind turbines equipped with doubly-fed induction generators, the initial rotor speed, $\omega_M(t=0)$, is found from the rotor initialisation set to the desired power output. When the desired power output is below the rated power, the rotor speed is found using Eq.(3.20). Otherwise the rotor speed is set to the rated rotor speed. The initial generator rotor speed in p.u., $\omega_G(t=0)$, is set equal to the initial rotor speed in p.u.

The initial twist, θ_S, will be:

$$\theta_S(t=0) = \frac{T_M(t=0) - D_M \cdot \omega_M(t=0)}{K_S} \ , \qquad (3.48)$$

when using:

$$T_M(t=0) = T_E(t=0) + D_M \cdot \omega_M(t=0) + D_G \cdot \omega_G(t=0). \qquad (3.48)$$

The electrical torque, T_E, is initialised from the generator model as shown in **Section 4.2.5**.

3.2.2.3. Identification of shaft system parameters

In earlier experiments, see (Pedersen et al., 2000), torsion oscillations with a significant magnitude were measured in the shaft systems of fixed-speed wind turbines. Torsion oscillations in the wind turbine shaft systems were confirmed by detailed experimental work performed by Sørensen et al. (2003) and Raben et al. (2003). In the experiment described by Pedersen et al. (2000), the Danish windfarm at Rejsby Hede, consisting of forty 600 kW wind turbines, was brought into is-land operation, e.g. tripping the windfarm from the local distribution power grid, during about one second. Pedersen et al. (2000) measured the grid frequency fluctuations and coupled such fluctua-tions with the torsion oscillations of the wind turbine shaft systems. The results of this experiment were used to evaluate the shaft system parameters (Akhmatov et al., 2003(a)) and also to explain the nature of the electromechanical interaction between wind turbines and the power grid (Pedersen et al., 2003).

Following the description of (Akhmatov et al., 2003(a)) and evaluating the shaft system parame-ters from the measurements of the grid frequency of the wind turbines brought into island operation. The shaft system parameters to be evaluated relate to the two-mass model by Eq.(3.46) using the rotor inertia constant, H_M, the generator rotor inertia constant, H_G, and the total shaft stiffness, K_S.

The experiment at Rejsby Hede windfarm was carried out when the wind speed was approxi-mately 10 m/s wind and the windfarm was approximately 80% reactive compensated. After discon-necting the windfarm from the grid, the fluctuating behaviour of the electric frequency was ob-served, see **Figure 3.18**.

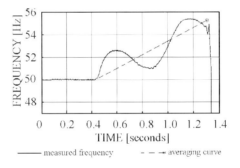

Figure 3.18   Electric frequency of the Rejsby Hede windfarm measured during electrical tripping experiment. Re-printed from Ref. (Pedersen et al., 2003), Copyright (2003), with permission from Elsevier.

To analyse the trend of the measured frequency curve, the movement equation of the lumped system is applied.

$$\begin{cases} \omega_L \cdot \dfrac{d\omega_L}{dt} = \dfrac{P_M}{2 \cdot (H_M + H_G)} \ , \\[3mm] \omega_L = \dfrac{H_M \cdot \omega_M + H_G \cdot \omega_G}{H_M + H_G}. \end{cases} \qquad (3.50)$$

where ω_L denotes the total speed of the lumped system consisting of the rotor and the generator rotor. The movement Eq.(3.50) is written for a disconnected wind turbine when the electrical torque, T_E, is zero. Disconnection occurs at time $t_1 = t_0$. Just before the disconnection, the generator rotor slip was $s(t_0)$ and the lumped system speed was therefore $\omega_L(t_1) = \omega_L(t_0)=1 - s(t_0)$. At time t_2

$=t_0 +\Delta t$, the lumped system speed became $\omega_L(t_2) = \omega_L(t_0) + \Delta\omega_L(t_0+\Delta t)$. The increase of the lumped system speed was found by integrating Eq.(3.50) from the time of tripping, $t_1 = t_0$, to the time t_2, thus:

$$\Delta\omega_L = \frac{\int_{t0}^{t0+\Delta t} P_M(\tau) \cdot d\tau}{2 \cdot \omega_L(t_0) \cdot (H_M + H_G)}.$$

(3.51)

With the use of Eq.(3.51), the rotor inertia constant is evaluated.

$$H_M = \frac{\int_{t0}^{t0+\Delta t} P_M(\tau) \cdot d\tau}{2 \cdot \omega_L(t_0) \cdot \Delta\omega_L} - H_G \approx \frac{\frac{1}{2} \cdot (P_M(t_0) + P_M(t_0 + \Delta t)) \cdot \Delta t}{2 \cdot \Delta f} - H_G.$$

(3.52)

Here Δf denotes the increase of the grid frequency that is proportional to the increase of the lumped system speed, $\Delta\omega_L$. The increase of the mechanical power of the rotor during the island operation, $P_M(t_0 + \Delta t) > P_M(t_0)$, is given by the coupling between the mechanical power, P_M, and the rotor speed, ω_M. This value is estimated from the slope for the C_P-λ-curve of the given wind turbine.

The generator inertia constant of the given wind turbine is set to $H_G = 0.5$ s. From the trend of the measured frequency curve shown in **Figure 3.18**, the parameters are estimated as $\Delta t = 0.85$ s, Δf =0.1 p.u. The expression containing the increase of the mechanical power of the rotor, $? \cdot (P_M(t_0) + P_M(t_0 + \Delta t))$, is evaluated to 0.70 p.u. With the use of Eq.(3.52), the rotor inertia constant is found $H_M = 2.5$ s. However, the oscillating behaviour of the measured grid frequency cannot be explained in terms of the lumped-mass model of the shaft system. With the use of a two-mass model Eq.(3.46), the movement equation of the shaft system of the tripped wind turbine is derived.

$$\frac{d^2}{dt^2}\theta_S + \frac{\omega_0 K_S}{2} \cdot \frac{H_M + H_G}{H_M H_G} \cdot \theta_S = 0 \ .$$

(3.53)

A homogenous solution of the movement Eq.(3.53) with regard to the shaft twist, θ_S, relates to the oscillating behaviour of the measured grid frequency, $f(t)$. When applying Eq.(3.53), the shaft torsion mode, f_T, is defined for such a homogenous solution as:

$$\frac{d^2}{dt^2}\theta_S + (2\pi f_T)^2 \cdot \theta_S = 0 \ .$$

(3.54)

The total shaft stiffness, K_S, is defined using Eq.(3.53) and Eq.(3.54).

$$K_S = \frac{8 \cdot \pi^2 \cdot f_T^2}{\omega_0} \cdot \frac{H_M \cdot H_G}{H_M + H_G} \ .$$

(3.55)

The period of the grid frequency oscillation is 0.6 s. The natural frequency f_T of such oscillations is therefore 1.7 Hz, which according to the results of (Pedersen et all, 2003) gives the shaft torsion mode. When applying Eq.(3.55) and values of H_G, H_M and f_T, the total shaft stiffness of a given medium-sized wind turbine is found as K_S =0.30 p.u./el.rad. This routine gives a general description of how the shaft system parameters can be evaluated from experiments.

To validate the shaft system parameters found using the above routine, the experiment at the Rejsby Hede windfarm was simulated for simplified conditions. The validation was performed in the case of a single wind turbine with the same parameters of the shaft system as found with the use of the above routine. The simulation case was executed at the same operational conditions of the wind turbine generator (i.e. wind speed is 10 m/s and induction generator is 80% reactive compensated). The simulated grid frequency is shown in **Figure 3.19**. The maxima and minima of the measured and the simulated curves of the grid frequency are compared in **Table 3.6**, and the measured and the simulated behaviours are in good agreement.

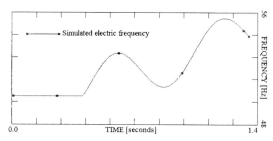

Figure 3.19 Simulated behaviour of grid frequency gained from the validation of the tripping experiment. Reprinted from Ref. (Akhmatov et al., 2003(a)), Copyright (2003), with permission from Elsevier.

Value	Measured	Simulated
First Maximum	53.0 Hz	53.0 Hz
First Minimum	51.0 Hz	50.6 Hz
Second Maximum	55.5 Hz	55.7 Hz
Period	0.6 s	0.6 s
Eigenfrequency	1.7 Hz	1.7 Hz

Table 3.6 Comparison between measured and simulated behaviour of grid frequency for the experiment. Reprinted from Ref. (Akhmatov et al., 2003(a)), Copyright (2003), with permission from Elsevier.

The results of the island experiment at Rejsby Hede windfarm demonstrated that the total shaft stiffness of modern wind turbines is significantly lower than the shaft stiffness of conventional power plants. It also demonstrated the nature of the electromechanical interaction between wind turbines and the power grid. In the case of fixed-speed wind turbines equipped with induction generators, this interaction occurs between the shaft systems of the turbines and the power grid and is caused by a strong coupling between the generator rotor slip and the electrical parameters of the induction generators (Akhmatov et all, 2000(a)).

3.2.2.4 Note on application

When compared to the lumped-mass model, the two-mass model appears more complex and re-quires more data. This must be seen in contrast with the representation of conventional power plants that are multi-shaft systems, but computed with the use of the lumped-mass model in investigations of short-term voltage stability.

Therefore it is necessary to evaluate where the two-mass model must be applied to represent wind turbines in investigations of short-term voltage stability and where the lumped-mass model can still be used. The torsion oscillations of the shaft systems represent internal swings in the me-chanical constructions of wind turbines. From this point of view, this behaviour might not necessar-ily relate to the grid operation and was disregarded in earlier investigations of short-term voltage stability (Bruntt et al., 1999). The use of the two-mass model to represent the shaft systems of wind turbines in investigations of short-term voltage stability is needed due to a significant electrome-chanical interaction between the shaft systems of the wind turbines and the power grid (Akhmatov et al., 2000(b)).

3.2.2.4.1 Fixed-speed wind turbines

In the case of fixed-speed wind turbines equipped with induction generators, there is a strong coupling between the generator rotor slip, s, and the electrical parameters of such induction genera-tors, such as the active power, P_E, reactive power, Q_E, and terminal voltage, U_S. Torsion oscillations of the shaft system are excited by grid disturbances, which then initiate fluctuations of the generator rotor speed as well as the electrical parameters of the generator. As explained in (Akhmatov et al., 2000(a)), the intensity of such fluctuations corresponds to the potential energy accumulated in the twisted shafts of the wind turbines at the pre-faulted, normal operation conditions of the power grid.

At normal operation of the power grid, the twisted shafts of the wind turbines accumulate an amount of the potential energy, W_S.

$$W_S = \tfrac{1}{2} \cdot K_S \cdot \theta_S^2 = \tfrac{1}{2} \cdot K_S \cdot \left(\frac{T_M}{K_S} \right)^2 = \tfrac{1}{2} \cdot \frac{T_M^2}{K_S} \, , \tag{3.56}$$

where θ_S denotes the initial twist at normal operation and damping, D_M and D_S, is disregarded. When the power grid is subject to a short-circuit fault, the terminal voltage, U_S, and then the electri-cal torque, T_E, of the wind turbine generator are reduced. The twisted shaft starts relaxing, and dur-ing the relaxation process, the potential energy of the shaft is transferred to kinetic energy in the generator rotor. This energy transformation process results in more acceleration of the generator rotor than would be predicted in the case of the lumped-mass model of shaft system (Akhmatov et al., 2000(b)). The reactive power absorption of the induction generators, Q_E, increases when the generator rotor speed and slip increase. Shaft relaxation then leads to absorption of more reactive power and slower voltage recovery in the grid than would be predicted with the use of the lumped-mass model. From this point of view, the use of the lumped-mass model of the wind turbine shafts would lead to a prediction of over-pessimistic voltage recovery rates in the power grid.

Though the two-mass model of the shaft systems represents an internal torsion swing in the shaft system, this shaft behaviour influences the voltage recovery rate in the power grid due to a strong coupling between the mechanical and the electrical parameters of the induction generators of fixed-speed wind turbines. This then gives the main explanation why the shaft system of fixed-speed wind turbines must be represented using two-mass models. The validity of the two-mass model is based on the fact that the total stiffness of the wind turbine shafts, K_S, is sufficiently low. This covers the major cases of medium-sized and large, modern wind turbines in the MW range.

3.2.2.4.2 Variable-speed wind turbines

The shaft stiffness of variable-speed wind turbines is in the same range as in the case of fixed-speed wind turbines. When the power grid is subject to a short-circuit fault, torsion oscillations are also present in the shaft system of variable-speed wind turbines. Such variable-speed wind turbines are equipped with doubly-fed induction generators controlled by frequency converters. The converter control is arranged with independent control of the active, P_E, and reactive power, Q_E, (Yamamoto and Motoyoshi, 1991). The excitation of doubly-fed induction generator is controlled by a frequency converter. Therefore the reactive power absorption and the grid voltage at the generator terminals are decoupled from the generator rotor speed. In such power conversion systems, torsion oscillations present in the shaft lead to fluctuations in the generator rotor speed, but do not affect the grid voltage at the generator terminals, U_S, so long as the frequency converter is in operation. The frequency converter contains power electronics components which are sensitive to electrical and thermal overloading. When the power grid is subject to a short-circuit fault, there is a risk that the frequency converter stops switching and blocks, to protect such power electronics components. The operation with a blocked converter may introduce a coupling between the generator rotor speed and the generator excitation. Such a coupling appears when the crow-bar protection of the frequency converter is applied (Petersson, 2003; Akhmatov, 2003(c)). The crow-bar protection of the frequency converters is discussed in **Section 7.7.1.2**. In such situations, shaft torsion oscillations affect the grid voltage at the generator terminals, similar to the case of fixed-speed wind turbines.

When the frequency converter is in operation (for example, when the crow-bar protection is removed and the converter is restarted), the active power, P_E, is set to follow the fluctuating behaviour of the generator rotor speed (Akhmatov, 2002(b)). This is required to dampen the shaft torsion oscillations excited at grid disturbances. The natural frequency of such power fluctuations is equal to the shaft torsion mode and derived from Eq.(3.55).

$$f_T = \frac{1}{2\pi} \sqrt{\frac{\omega_0 \cdot K_S \cdot (H_M + H_G)}{2 \cdot H_M \cdot H_G}} . \tag{3.56}$$

The shaft torsion mode of wind turbines is in the range of 1 to 2 Hz which is relatively close to the natural frequencies of the synchronous generators of conventional power plants. There is a risk of torsion oscillations exciting corresponding oscillations in the synchronous generators of conventional power plants in situations with insufficient damping. When the stiffness of the shaft in variable-speed wind turbines is sufficiently low, there is a good reason to use the two-mass shaft system to represent the wind turbine shafts in investigations of power system stability.

3.2.2.4.3 Range of lumped-mass model

In situations when the shaft system of wind turbines is sufficiently stiff, ideally when $K_S \rightarrow \infty$, the lumped-mass model is used without loss of accuracy. The potential energy, W_S, accumulated in the twisted shafts is reciprocal to the shaft stiffness, K_S. When the shaft is ideally stiff, it does not twist and then does not accumulate potential energy. Akhmatov and Knudsen (2002) reported that when the shaft stiffness K_S =3.0 p.u./el.rad., there are neither significant fluctuations in the genera-tor rotor speed nor in other electrical parameters of the wind turbine. In this case, the shaft torsion oscillations are seen as ripples with a very small magnitude and do not affect the voltage in the power grid.

This value of the shaft stiffness, i.e. K_S =3.0 p.u./el.rad., may define the boundary for the validity of the two-mass model. When the shaft stiffness, K_S, is lower than 3.0 p.u./el.rad., the two-mass model must be applied to represent the shaft systems of electricity-producing wind turbines. When the shaft stiffness is equal or larger than 3.0 p.u./el.rad., the lumped-mass model can be applied. In the case of the lumped-mass model, the shaft stiffness, K_S, is set to infinity and the shaft twist, θ_S, is set to zero. In terms of the lumped-mass model, the rotor inertia and the generator rotor inertia are lumped together to form a single rotating mass, H_L =H_M +H_G.

3.3. Summary

A modern wind turbine is a complex mechanical construction. In investigations of power system stability, representations of the mechanical construction can be reduced to a shaft system model, an aerodynamic rotor model, and a blade angle control.

The shaft system is essentially the two-mass model (Akhmatov et al., 2000(a)). In the two-mass model, the rotor and the generator rotor are connected through a shaft with relatively slow stiffness. The two-mass model computes the shaft torsion oscillations viewed from the terminals of the induc-tion generator. The shaft torsion oscillations lead to fluctuations of the generator rotor speed. Due to a strong coupling between the generator rotor speed and the electrical parameters of the generator, such fluctuations will also occur in the electrical parameters of the generator. Then, the shaft torsion oscillations may affect operation of the generator and the power grid.

The natural frequency of the shaft torsion oscillations is in the range of 1 to 2 Hz indicating a relatively soft coupling between the rotor and the generator rotor. In the case of fixed-speed wind turbines with induction generators, the shaft torsion oscillations lead to slowing of the grid voltage recovery after a grid fault (Akhmatov et al., 2000(a)). In the case of variable-speed wind turbines with doubly-fed induction generators, the shaft torsion oscillations may introduce interaction be-tween the wind turbine generators and the synchronous generators of conventional power plants.

When the shaft stiffness becomes sufficiently large, the lumped-mass model can be used to rep-resent the rotating parts of the wind turbine construction. In the lumped-mass model, the shaft stiff-ness is assumed to be equal to infinity. The rotor mass and the generator rotor mass are then lumped together into a single rotating mass. When the shaft stiffness is equal to or larger then 3.0 p.u./el.rad., the lumped-mass model can be applied instead of the two-mass model.

The aerodynamic rotor model gives the coupling between the rotor speed and the mechanical power of the rotor. The mechanical power of the rotor can be computed in several ways depending on the desired accuracy and the target of investigations.

The pitch control model gives the dynamic coupling between the pitch angle and the mechanical power of the rotor. The pitch angle is controlled with the use of the blade-angle control system setting the pitch angle reference. The pitch servo adjusts the pitch angle to its reference value.

In investigations of power system stability, the mechanical power of the rotor can be computed on the assumption that the rotor is always in equilibrium. Then, the stationary methods such as the use of pre-computed $C_P(\lambda,\beta)$ characteristics or the BEM method can be applied to compute the mechanical power (and the torque) of the rotor. When pitch control is used, this assumption is valid at slow pitching. When the active-stall control is applied, the aerodynamic rotor model based on this assumption will predict accurate results also.

To improve the accuracy of the computations at fast pitching, the rotor model with representation of dynamic inflow phenomena may be required. This improved model predicts notable overshoots appearing at fast pitching in the mechanical torque of the rotor of the pitch controlled wind turbines. The magnitude of the overshoots depends on operational conditions of the rotor and the rotor radius. The magnitude of the overshoots increases at increasing rotor radius.

In investigations of power system stability, the mechanical power (and the torque) of the rotor, as well as the states of the shaft system model are initialised from the generator initialisation.

The mechanical power (and the torque) of the rotor and the states of the shaft system model may bee computed in the p.u. system. Then, the p.u. system of the mechanical construction model is defined with regard to the p.u. system of the generator.

Energy service at Utgrunden Windfarm. Photo copyright GE Wind Energy. Reproduced with permission from GE Wind Energy.

4 Induction generator models

Induction generators with a short-circuited rotor circuit are applied in fixed-speed wind turbines to convert the mechanical shaft power into electrical power. The stator of the induction generator is placed on the top of the wind turbine tower and according to a common assumption on the induction generator modelling stands still (Krause, 1995; Kundur, 1994). The generator rotor is connected to the high-speed shaft of the drive-train system and rotates with the mechanical speed, ω_R.

In induction generators, the active power, P_E, is supplied to the power grid whereas the reactive power, Q_E, is absorbed from the power grid to magnetise the generator. The induction generators have no excitation control, but are excited from the power grid. This implies that induction generators cannot control the grid voltage. The grid connection requires that the grid voltage at the induction generator terminals, U_S, is maintained by control arrangements of the power grid and in the range of the rated voltage of the induction generator, U_{RAT}.

The active and the reactive power of the induction generators depend on the generator rotor slip, i.e. the relative deviation of the generator rotor speed, $\omega_G = N_{EE}\omega_R$, from the electrical grid speed, ω_E.

$$s = \frac{\omega_E - \omega_G}{\omega_E} = \frac{\omega_E - N_{EE} \cdot \omega_R}{\omega_E}. \tag{4.1}$$

N_{EE} denotes the number of pole-pairs of the induction generator. Notice that ω_G is the generator rotor speed computed with regard to the electrical grid speed. In the generator operation mode, the generator rotor speed, ω_G, exceeds the electrical grid speed, ω_E. Therefore the slip, s, is negative in generator mode. In contrast, the slip is positive in motor mode. When the grid frequency is f_E, the electrical grid speed is given as:

$$\omega_E = 2\pi \cdot f_E. \tag{4.2}$$

In classical power systems, the grid frequency is kept (almost) fixed at the rated grid frequency, f_{RAT}, using control of conventional power plants. The mechanical rotor speed of synchronous generators is kept at $2\pi f_{RAT}/N_{EE}$ where N_{EE} is the number of pole-pairs of the generators.

Existing simulation tools applicable for investigations of power system stability contain models of induction generators. Models of induction generators and their respective considerations are already well-described in for example Ref. (Krause, 1995; Kundur, 1994, Heier, 1998). This text meanwhile gives an overview of induction generator models of different orders with a focus on their application to investigations of power system stability.

4.1 Per unit system

The per unit system applied for modelling induction generators must comply with the per unit system applied for the network solution as well as with the per unit system applied to mechanical system models. All the parameters of the induction generator model are computed with regard to the base- values:

$$Y(p.u.) = Y(physical\ units)/Y_{BASE}(physical\ units),\tag{4.3}$$

where such physical units can be Volts, Amps, Watts, etc., Y denotes a given parameter of the induction generator model. The selected base- values are listed below:

$S_{BASE} = S_{RAT}$ is the MVA-base, e.g. the rated apparent power to compute the active, P_E, and the reactive power, Q_E, in p.u.

$f_{BASE} = f_{RAT}$ is the base- frequency of the grid frequency, f_E.

$U_{BASE} = U_{RAT}$ is the base- voltage of the stator voltage, U_S, and of the rotor voltage in stator units, U_R.

$I_{BASE} = I_{RAT}$ is the base- current of the stator current, I_S, and the rotor current in stator units, I_R.

$\psi_{BASE} = U_{RAT}/(2\pi \cdot f_{RAT})$ is the base- flux of the stator flux, ψ_S, and the rotor flux in stator units, ψ_R.

$Z_{BASE} = U_{RAT}/I_{RAT}$ is the base- impedance of the resistance and the reactance of the induction generator circuits in stator units.

$t_{BASE} = 1/\omega_{BASE} = 1/(2\pi \cdot f_{RAT})$ is the base- time.

4.2 State equations of induction generators

The state equations of induction generators transformed to a synchronously rotating reference frame (the D- and Q- quantities) are:

$$\begin{cases} u_{DS} = R_S \cdot i_{DS} - \dfrac{\omega_E}{\omega_{BASE}} \cdot \psi_{QS} + \dfrac{1}{\omega_{BASE}} \cdot \dfrac{d\psi_{DS}}{dt}, \\[2mm] u_{QS} = R_S \cdot i_{QS} + \dfrac{\omega_E}{\omega_{BASE}} \cdot \psi_{DS} + \dfrac{1}{\omega_{BASE}} \cdot \dfrac{d\psi_{QS}}{dt}, \\[2mm] u_{DR} = R_R \cdot i_{DR} - \dfrac{1}{\omega_{BASE}} \cdot \dfrac{d\theta_R}{dt} \cdot \psi_{QR} + \dfrac{1}{\omega_{BASE}} \cdot \dfrac{d\psi_{DR}}{dt}, \\[2mm] u_{QR} = R_R \cdot i_{QR} + \dfrac{1}{\omega_{BASE}} \cdot \dfrac{d\theta_R}{dt} \cdot \psi_{DR} + \dfrac{1}{\omega_{BASE}} \cdot \dfrac{d\psi_{QR}}{dt}. \end{cases} \tag{4.4}$$

The parameters of the state equations are all in p.u. accordingly to Eq.(4.3) whereas the electrical system speed, ω_E, is in el.rad./s and the rotor angle, θ_R, is in el.rad. The terminal voltage is $U_S = u_{DS} + j u_{QS}$ and the stator current is $I_S = i_{DS} + j i_{QS}$. The rotor parameters, including the rotor current, $I_R = i_{DR} + j i_{QR}$, are all in the stator windings. The rotor angle, θ_R, is the integral over time of the electromagnetic field speed in the rotor. In general, the states of the induction generator model, ψ_{DS}, ψ_{QS}, ψ_{DR} and ψ_{QR}, are fluxes and rotor angle, θ_R.

When the induction generator is with a short-circuited rotor circuit, the rotor voltage is a zero-vector:

$$U_R = \begin{bmatrix} u_{DR} \\ u_{QR} \end{bmatrix} = \begin{bmatrix} 0 \\ 0 \end{bmatrix}. \tag{4.5}$$

The derivatives of the states, $d\psi_{DS}/dt$, $d\psi_{QS}/dt$, $d\psi_{DR}/dt$ and $d\psi_{QR}/dt$ are all zero when the induction generator is in stationary operation, except for the derivative of the rotor angle, $d\theta_R/dt$. In steady-state conditions, the derivative of the rotor angle is equal to the stationary speed of the electromagnetic field in the rotor. The derivative of this angle is:

$$\frac{d\theta_R}{dt} = s \cdot \omega_E, \tag{4.6}$$

where the electrical system speed is given by Eq.(1.2). When Eq.(4.6) is inserted into Eq.(4.4), the rotor angle, θ_R, is excluded from the states of the induction generator model.

When the movement equation for the generator rotor speed, ω_G, is added to the state-equation system Eq.(4.4), the system contains five coupled differential equations for the five states ψ_{DS}, ψ_{QS}, ψ_{DR}, ψ_{QR} and ω_G. The state-equation system is called a transient, fifth-order model of induction generators. This transient, fifth-order model, acknowledges the transients in the rotor circuit as well as the fundamental-frequency transients in the stator flux, $\psi_S = [\psi_{DS}, \psi_{QS}]^T$ and in the stator current $I_S = [i_{DS}, i_{QS}]^T$. The term "fundamental-frequency" implies that the stator transients appear with the fundamental of the grid frequency that is the rated grid frequency, f_{RAT}. To compute the fundamental-frequency transients with a sufficient resolution, the time step might be taken to be one-fifth of a half-period of such fundamental-frequency transients in the stator flux and the stator current. Then the time step is about 2 ms in the case of a 50 Hz system.

4.2.1 Equations of flux linkages

Figure 4.1 presents the coupling between the flux linkages and the current. By inspection of **Figure 4.1**, the expressions for the flux linkages become:

$$\begin{cases} \psi_S = (X_S + X_M) \cdot I_S + X_M \cdot I_R, \\ \psi_R = (X_R + X_M) \cdot I_R + X_M \cdot I_S. \end{cases} \tag{4.7}$$

Here X_S is the reactance of the stator circuit, X_R is the reactance of the rotor circuit and X_M denotes the magnetising reactance.

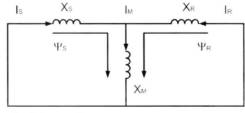

Figure 4.1 Coupling between flux linkages and currents.

When applying notation of the flux linkages and the currents, the flux linkages will be:

$$\begin{cases} \psi_{DS} = (X_S + X_M) \cdot i_{DS} + X_M \cdot i_{DR}, \\ \psi_{QS} = (X_S + X_M) \cdot i_{QS} + X_M \cdot i_{QR}, \\ \psi_{DR} = (X_R + X_M) \cdot i_{DR} + X_M \cdot i_{DS}, \\ \psi_{QR} = (X_R + X_M) \cdot i_{QR} + X_M \cdot i_{QS}. \end{cases} \tag{4.8}$$

Note that Eq.(4.8) are applied together with the state equations (4.4) as an iterative routine to perform computations of the transient, fifth-order model of induction generators at each time step.

4.2.2 Reduced third-order model

The reduced, third-order model of induction generators is based on the representation of the induction generator by the transient voltage source, $E' = e'_D + j e'_Q$, behind the transient impedance, $Z' = R_S + jX'$. This representation is illustrated in **Figure 4.2**.

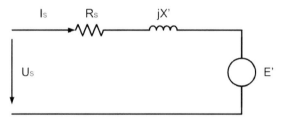

Figure 4.2　Transient equivalent of an induction generator with a single rotor circuit.

The real part of the transient impedance is the stator resistance, R_S, and the imaginary part is the transient reactance, X', that is:

$$X' = X_S + X_M \parallel X_R = X_S + \frac{X_M \cdot X_R}{X_M + X_R}. \tag{4.9}$$

The transient voltage source, E', can be written in the terms of the rotor flux, ψ_R, (Kundur, 1994; Feijóo et al., 2000):

$$\begin{cases} e'_D = \dfrac{-\omega_E}{\omega_{BASE}} \cdot \dfrac{X_M}{X_M + X_R} \cdot \psi_{QR}, \\ e'_Q = \dfrac{\omega_E}{\omega_{BASE}} \cdot \dfrac{X_M}{X_M + X_R} \cdot \psi_{DR}. \end{cases} \tag{4.10}$$

The components of the transient voltage source together with the generator rotor speed, ω_G, are often applied as the states of a reduced, third-order model of induction generators. These state equations are gained from the state equations of the transient fifth-order model Eq.(4.4) when disregarding the stator flux transients:

$$\begin{cases} d\psi_{DS}/dt = 0, \\ d\psi_{QS}/dt = 0. \end{cases} \tag{4.11}$$

Applying Eq.(4.11), it can be shown that equations for the reduced, third-order model of induction generators are:

$$\begin{cases} \dfrac{de'_D}{dt} = -\dfrac{\omega_E \cdot R_R}{X_R + X_M} \cdot \left(e'_D + \left(\dfrac{X_M^2}{X_M + X_R} \right) \cdot i_{QS} \right) + s \cdot \omega_E \cdot e'_Q, \\[3mm] \dfrac{de'_Q}{dt} = -\dfrac{\omega_E \cdot R_R}{X_R + X_M} \cdot \left(e'_Q - \left(\dfrac{X_M^2}{X_M + X_R} \right) \cdot i_{DS} \right) - s \cdot \omega_E \cdot e'_D, \\[3mm] u_{DS} = R_S \cdot i_{DS} - X' \cdot i_{QS} + e'_D, \\[2mm] u_{QS} = R_S \cdot i_{QS} + X' \cdot i_{DS} + e'_Q. \end{cases} \tag{4.12}$$

Similar expressions may be found in (Kundur, 1994) and (Feijóo et al., 2000). When the stator flux and current transients are disregarded, the time step of the model may be increased significantly. Usually, the time step of 10 ms may be applied to compute the model of induction generators as well as the power network model.

4.2.3 Relevant generator model parameters

In literature dedicated to electric machinery, for example, (Kundur, 1994; Krause, 1995), the state equations for the induction generator are written with the use of other electrical parameters computed from R_S, X_S, X_M, R_R and X_R. These parameters are given in **Table 4.1**.

The parameters collected in **Table 4.1** may also be required as the input data for the induction generator models in the simulation tools applied for investigations of short-term voltage stability, for example in the tool PSS/E^{TM}.

Parameters	Expressions
Per unit stator winding leakage inductance	$L_S = X_S$ at $f_E = 1.0$ p.u.
Per unit rotor winding leakage inductance	$L_R = X_R$ at $f_E = 1.0$ p.u.
Self-inductance of stator winding	$L_{SS} = L_S + L_M$
Self-inductance of rotor winding	$L_{RR} = L_R + L_M$
Transient reactance of a single-cage induction generator	$X' = X_S + (1/X_M + 1/X_R)^{-1}$
Transient open-circuit time constant expressed in seconds of a single-cage induction generator	$T'_0 = (X_M + X_R)/(\omega_{BASE} R_R)$

Table 4.1 Definition of electrical parameters relevant for induction generator modelling.

4.2.4 Electrical power

The apparent power of induction generators is expressed as:

$$\begin{cases} S_E = P_E + j \cdot Q_E = U_S \cdot conj(I_S), \\ P_E = u_{DS} \cdot i_{DS} + u_{QS} \cdot i_{QS}, \\ Q_E = u_{DS} \cdot i_{QS} - u_{QS} \cdot i_{DS}, \end{cases} \tag{4.13}$$

where S_E is the apparent power, P_E is the active power and Q_E is the reactive power of induction generators.

4.2.5 Electrical torque

The electrical torque of the generator, T_E, appears in the movement equation of the generator rotor and in the shaft system equations Eq.(3.46). This is the decelerating torque reducing the generator rotor acceleration during grid disturbances. In terms of the transient, fifth-order model, the electrical torque is expressed as:

$$T_E = \psi_{QR} \cdot i_{DR} - \psi_{DR} \cdot i_{QR}. \tag{4.14}$$

In terms of the reduced, third-order model, the electrical torque is found using:

$$T_E = e'_D \cdot i_{DS} + e'_Q \cdot i_{QS}. \tag{4.15}$$

4.2.6 Steady-state equations

Steady-state modelling corresponds to initialisation of the induction generator model prior to execution of dynamic simulations. Under initialisation, the slip, the active and the reactive power, the stator and the rotor current and the states are found in the given operation situation in steady-state. The derivatives of the states are all zero.

In steady-state, the grid frequency is equal to the rated grid frequency. Then, the transformation of the equations is made with regard to the synchronously rotating reference frame with the speed of the transformation $\omega_T = 2\pi f_{RAT}$. The electrical parameters are again in p.u. **Figure 4.3** shows the equivalent circuit of the induction generator with a short-circuited rotor circuit in steady-state. Since the induction generator is symmetrical in all the three physical phases, then the electrical equivalents for the D-axis and for the Q-axis are identical. Therefore only a single equivalent is applied to compute the D- and the Q- quantities.

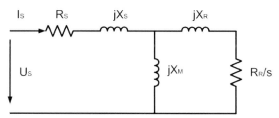

Figure 4.3 Equivalent circuit of an induction generator with a squirrel-cage rotor circuit.

The impedance of the equivalent circuit viewed from the induction generator terminals is given by:

$$Z_T = R_T + jX_T = (R_s + j \cdot X_s) + j \cdot X_M \left\| \left(\frac{R_R}{s} + j \cdot X_R \right) \right. \tag{4.16}$$

Note that the resistance, R_T, and the reactance, X_T, of the equivalent circuit impedance are slip- dependent:

$$\begin{cases} R_T(s) = R_S + \dfrac{(R_R/(-s)) \cdot X_M^2}{(R_R/s)^2 + (X_M + X_R)^2} \quad , \\[4mm] X_T(s) = X_S + \dfrac{X_M \cdot \left((R_R/s)^2 + X_R \cdot (X_M + X_R) \right)}{(R_R/s)^2 + (X_M + X_R)^2} \quad . \end{cases} \tag{4.17}$$

The apparent power of the induction generator in steady-state becomes:

$$S_E = P_E + j \cdot Q_E = U_S \cdot conj(I_S) = U_S \cdot conj\left(\frac{U_S}{Z_T} \right) = |U_S|^2 \cdot \left(\frac{R_T + j \cdot X_T}{R_T^2 + X_T^2} \right). \tag{4.18}$$

Then the active power supplied from, and the reactive power absorbed by, the induction generator is:

$$\begin{cases} P_E = |U_S|^2 \cdot \dfrac{R_T}{R_T^2 + X_T^2}, \\[4mm] Q_E = |U_S|^2 \cdot \dfrac{X_T}{R_T^2 + X_T^2}, \end{cases} \tag{4.19}$$

which are dependent on the terminal voltage magnitude and the generator rotor slip.

4.2.6.1 Initialisation of slip and reactive power

Initialisation is based on the active power, P_E, being known (set by the user at the beginning of the load-flow computation and fixed during initialisation) and the grid voltage magnitude, $|U_S|$, found from the network solution performed using a simulation tool. The initialisation routine is then applied to initialise the generator rotor slip, s, and the reactive power, Q_E, of the induction genera-tor.

As a first step, the generator rotor slip, s, is adjusted until the desired active power, P_E, is reached at a fixed voltage magnitude, $|U_S|$. In this adjusting routine, equations Eq.(4.16) to Eq.(4.19) are used. The reactive power absorbed by the induction generator, Q_E, can then be computed for the given slip, s, and the given voltage magnitude, $|U_S|$, by using of Eq.(4.19).

The network solution is then found for the fixed active power, P_E, and the reactive power, Q_E, from the prior step. The network solution results in an adjusted value of grid voltage magnitude, $|U_S|$. Then, going back to the first iterative step and applying the recently adjusted solution for the grid voltage magnitude, $|U_S|$. This iterative loop continues until the required accuracy of the slip and the grid voltage is attained.

4.2.6.2 Initialisation with a shunt

In existing power system stability simulation tools, the desired values for the active power, P_E, and the reactive power exchanged between the induction generator and the power grid, Q_T, can be established by the user. In this case, the values of P_E and Q_T are fixed input parameters to the network solution, which results in the grid voltage magnitude at the generator terminals, $|U_S|$.

For the next step, the generator rotor slip, s, is found by applying equations Eq.(4.16) to Eq.(4.19) at fixed values of active power, P_E, and grid voltage magnitude, $|U_S|$. When the slip is found, the reactive power absorption of the induction generator, Q_E, is computed from Eq.(4.19).

Finally, the simulation tool inserts a shunt at the induction generator terminals. The reactive power of the shunt, Q_{SH}, is the difference between the exchanged reactive power, Q_T, and the reactive power absorbed by the induction generator, Q_E.

$$Q_{SH} = Q_T - Q_E(s, U_S) . \tag{4.20}$$

4.2.6.3 Initialisation of current, flux and torque

When the grid voltage at the generator terminals, $U_S = u_{DS} + j\, u_{QS}$, is known from the network solution and the apparent power of the induction generator is initialised, the stator current is computed using Eq.(4.13).

$$\begin{cases} I_S = i_{DS} + j \cdot i_{QS}, \\[2mm] i_{DS} = \dfrac{P_E \cdot u_{DS} + Q_E \cdot u_{QS}}{u_{DS}^2 + u_{QS}^2}, \\[3mm] i_{QS} = \dfrac{P_E \cdot u_{QS} - Q_E \cdot u_{DS}}{u_{DS}^2 + u_{QS}^2} . \end{cases} \tag{4.21}$$

The transient voltage source, E', is initialised knowing the grid voltage, U_S, and the stator current, I_S, using Eq.(4.12):

$$\begin{cases} e'_D = u_{DS} - R_s \cdot i_{DS} + X' \cdot i_{QS}, \\ e'_Q = u_{QS} - R_s \cdot i_{QS} - X' \cdot i_{DS}. \end{cases} \quad (4.22)$$

The fluxes in the stator and the rotor circuit are initialised using Eq.(4.10) and Eq.(4.4) with the derivatives of the stator flux set to zero:

$$\begin{cases} \psi'_{DR} = \dfrac{X_M + X_R}{X_M} \cdot e'_Q, \\ \psi'_{QR} = -\dfrac{X_M + X_R}{X_M} \cdot e'_D, \end{cases}$$
$$\begin{cases} \psi_{DS} = u_{QS} - R_s \cdot i_{QS}, \\ \psi_{QS} = -u_{DS} + R_s \cdot i_{DS}. \end{cases} \quad (4.23)$$

The rotor current, I_R, is initialised from Eq.(4.8) when the rotor flux and the stator current are known:

$$\begin{aligned} i_{DR} &= \frac{\psi'_{DR} - X_M \cdot i_{DS}}{X_M + X_R}, \\ i_{QR} &= \frac{\psi'_{QR} - X_M \cdot i_{QS}}{X_M + X_R}. \end{aligned} \quad (4.24)$$

Finally, the electrical torque, T_E, is initialised with the use of either Eq.(4.14) or Eq.(4.15). Then, the shaft system is initialised applying Eq.(3.48) and Eq.(3.49) knowing the generator rotor slip and electrical torque in steady-state conditions. The steady-state value of the active power is applied in computations of the mechanical power of the rotor, P_M, according to the description in **Section 3.1.8**.

4.2.7 Representation of saturation

So far the effects of stator and rotor iron saturation have been disregarded. Based on this assumption, the relation between the flux and current for a given winding is given by the linear air-gap characteristic shown in **Figure 4.4**. Saturation represents non-linearity of the relation between the flux and current through the winding at excessive currents. In stability investigations, the saturation can be modelled by simplified, semi-empirical open-circuit characteristics, and by making some assumptions (Kundur, 1994). There are several mathematical ways to model the saturation with the use of such characteristics. These are illustrated in **Figure 4.4**. The saturation effects are modelled with the use of the saturation factor, K_{SAT}, which identifies the degree of saturation in the given winding.

$$\begin{aligned} X_{SAT} &= K_{SAT} \cdot X_{USAT}, \\ \psi &= X_{SAT} \cdot I, \end{aligned} \quad (4.25)$$

where X_{USAT} and X_{SAT} are the unsaturated reactance and the saturated reactance of the winding, respectively, ψ is the flux linkage and I is the current through the winding. The saturation factor, K_{SAT}, of the operational point (A,ψ_1) shown in **Figure 4.4** is defined as:

$$K_{SAT} = \frac{B}{A}, \qquad (4.26)$$

and then by substituting for A and B, as:

$$K_{SAT} = \frac{\psi_1}{\psi_1 + \Delta\psi} = \frac{\psi_1}{\psi_2}, \qquad (4.27)$$

where the values of the flux linkage, ψ_1 and ψ_2, are marked in **Figure 4.4**. The flux linkage ψ_{SAT} denotes the value where the saturation has just started and is also marked in **Figure 4.4**. Accordingly to (Kundur, 1994), the flux linkage difference, $\Delta\psi$, can be expressed by the exponential function:

$$\Delta\psi = \psi_2 - \psi_1 = A_{SAT} \cdot \exp\big(B_{SAT} \cdot (\psi_1 - \psi_{SAT})\big). \qquad (4.28)$$

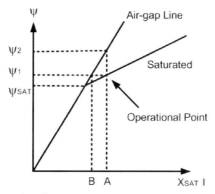

Figure 4.4   Representation of saturation effects.

Then, the saturation factor at the operational point (A,ψ_1) is defined without use of ψ_2, i.e.:

$$K_{SAT} = \frac{\psi_1}{\psi_1 + A_{SAT} \cdot \exp\big(B_{SAT} \cdot (\psi_1 - \psi_{SAT})\big)}, \qquad (4.29)$$

with the estimated parameters A_{SAT}, B_{SAT} and ψ_{SAT} of the given open-circuit characteristic. Alternatively, saturation may be represented using saturation parameters S defined as:

$$S = \frac{A - B}{B} = \frac{1}{K_{SAT}} - 1. \qquad (4.30)$$

Specifically, saturation parameters $S(1.0)$ and $S(1.2)$ for the flux linkages of $E'=1.0$ and $E'=1.2$ are applied in the simulation tool PSS/E[TM] to compute the saturation effects of generators. The parameters $S(1.0)$ and $S(1.2)$ are coupled with the parameters A_{SAT}, B_{SAT} and ψ_{SAT}, thus:

$$S = \frac{A - B}{B} = \frac{\psi_2 - \psi_1}{\psi_1} = \frac{A_{SAT} \cdot \exp(B_{SAT} \cdot (\psi_1 - \psi_{SAT}))}{\psi_1}. \tag{4.31}$$

Therefore these saturation parameters become:

$$\begin{cases} S(1.0) = A_{SAT} \cdot \exp(B_{SAT} \cdot (1.0 - \psi_{SAT})), \\ S(1.2) = \dfrac{A_{SAT} \cdot \exp(B_{SAT} \cdot (1.2 - \psi_{SAT}))}{1.2}. \end{cases} \tag{4.32}$$

The influence of saturation effects on the results of short-term voltage stability will be discussed in **Section 4.3.2.3**. Saturation effects play a role in the generation of higher harmonics of the fundamental frequency as these represent nonlinearity (Bolik, 2004). When the transient fifth-order model of induction generators includes representation of saturation, this may require further reduction of the time step in the computation.

4.2.8 Typical generator data

Electrical and mechanical data of induction generators are part of the design parameters. However, it is possible to introduce the order of such parameters. **Table 4.2** gives common data for induction generators in the range of tens of kW up to several MW. Note that the rated default voltage of induction generators up to 2.5 MW used in fixed-speed wind turbines is 0.69 kV (50 Hz systems). The rated voltage of wind turbine induction generators up to 3 MW and larger may be equal to or exceed 1 kV. For example, the generator rated voltage in the V90 3 MW wind turbine from Vestas Wind Systems is 1 kV. Vestas has also announced that the V120 4.5 MW wind turbine has a generator with a rated voltage of 6 kV.

Rated power, kW	H_G, s	R_S, p.u.	X_S, p.u.	X_M, p.u.	R_R, p.u.	X_R, p.u.
$10 < P_{RAT} < 100$	0.25	0.04	0.14	3.4	0.035	0.12
$100 < P_{RAT} < 1000$	0.5	0.015	0.12	4	0.015	0.15
$1000 < P_{RAT}$	0.7	0.01	0.12	5	0.005	0.2

Table 4.2 Order of electrical and mechanical parameters of induction generators.

The behaviour of induction generators subject to a short-circuited fault and the outcome of stability investigations are sensitive to generator parameters. This issue is specifically treated in **Section 5.3.2**. Apply therefore the exact data given by the wind turbine manufacturer, when possible.

4.3   Models for stability investigations

This section compares the results relating to voltage stability investigations that are reached with the use of the transient fifth-order model against the reduced third-order model of induction generators. The influence of saturation on model accuracy and on the results of voltage stability investigations will also be briefly discussed, focusing on the parameters of voltage stability investigation such as the recovery rate of grid voltage, excessive overspeeding of the generator rotor and excessive transients of the generator current. Such parameters relate to power system stability and fault ride-through capability of the wind turbines, and that the predicted outcome of the investigations may be dependent on the kind of model of induction generators used (Akhmatov and Knudsen, 1999).

4.3.1   Types of grid faults

Investigations of short-term voltage stability are normally carried out for power grids subject to short-circuit faults of a specific kind and of a specific duration. It is relevant to know the value of the voltage drop being caused by different types of short-circuit faults (such as 3-phase, 2-phase-to-ground, phase-to-phase and single-phase-to-ground short-circuit faults). **Table 4.3** gives typical values of the residual positive-sequence voltage at the faulted node subject to different types of short-circuit faults. This information is relevant when investigations of short-term voltage stability are carried out with simulation tools which apply positive-sequence equivalents of power grids, for example, the simulation tool PSS/E^{TM}.

Fault types	3-phase	2-phase-to-ground	Phase-to-phase	Single phase-to-ground
Residual positive-sequence voltage, p.u.	Less than 0.02	0.4	0.5	0.75

Table 4.3   Residual positive-sequence voltage in the faulted node at different types of short-circuit faults with zero-impedance.

4.3.2   Significance of current transients

Commonly in investigations of short-term voltage stability, induction generators have been represented using the reduced, third-order model (Kundur, 1994). This consideration is based on the experience with representation of synchronous generators in such investigations and coupling between the rotor angle and the dynamic stability of the synchronous generators (Kundur, 1994). When the dynamic stability relates to the behaviour of the rotor angle, which is the integral of the generator rotor speed, the average behaviour is more significant than the instantaneous behaviour of the speed. The electrical parameters of induction generators are however strongly coupled to the generator rotor slip (the speed deviation). The instantaneous behaviour of the generator rotor speed will therefore influence the reactive power absorption as well as the voltage behaviour of induction generators. **Figure 4.5** compares the generator terminal voltage, the speed and the current of induction generators modelled using the reduced, third-order model against the transient, fifth-order model. The simulations have been made for a 2 MW induction generator, with a 3-phase short-circuit fault as this kind of a fault gives the largest voltage drop. In this simulation, the faulted node is displaced by two transformers from the generator terminals and lasts 100 ms. The fault is then removed.

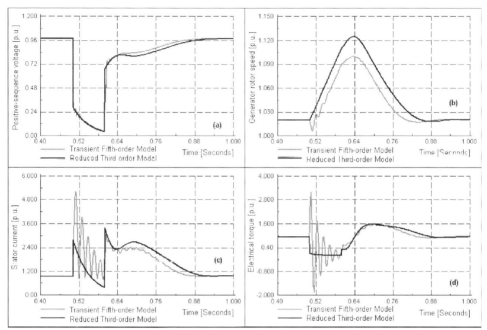

Figure 4.5 Comparison between the transient fifth-order model and the reduced third-order model of induction generators: **(a)** - terminal voltage, **(b)** - generator rotor speed, **(c)** - stator current, **(d)** - electrical torque.

The voltage recovery rate is dependent on the instantaneous generator rotor speed. The reduced third-order model may predict results that are pessimistic with regard to the generator rotor over-speeding and with regard to the voltage re-establishment after a grid fault. This result is predicted as the reduced third-order model disregards the fundamental-frequency transients in the stator flux.

$$\begin{cases} d\psi_{DS} / dt = 0, \\ d\psi_{QS} / dt = 0. \end{cases} \tag{4.33}$$

The result of this model simplification is seen in the generator rotor speed behaviour at the moment of the fault occuring. The generator rotor speed in the transient, fifth-order model enters the grid fault with a notch - the short-term reduction of the speed that is caused by the generator braking torque (Akhmatov and Knudsen, 1999). The electrical torque of the induction generator is plotted in **Figure 4.5** for the case of a 3-phase short-circuit fault. However, the generator rotor speed predicted by the reduced third-order model starts to increase immediately when the fault occurs. This behaviour is predicted because the reduced, third-order model disregards the braking torque at the moment of fault occurrence. The result of this model-dependent discrepancy is that the reduced third-order model predicts more overspeeding and then larger demands of reactive power to excite the induction generator when the fault is cleared. This discrepancy in generator rotor speed at the moment of the fault results in the prediction of slower voltage recovery when the reduced third-order model is applied. Moreover, the stator current predicted by the reduced, third-order model does not contain fundamental-frequency transients, and its magnitude is notably lower than the current magnitude in the transient, fifth-order model. This discrepancy is caused as disregarding of the stator flux transients corresponds to disregarding of the DC-offset in the generator current during balanced symmetrical events in the grid. The model-related difference in the machine current magnitude can be between 30 to 40%.

Figure 4.6 gives the single phase currents of the induction generator computed with the use of the transient, fifth-order model compared with the reduced, third-order model. In this computation, the 3-phase short-circuit fault is applied to the generator terminals and not removed to see how long it would take for the generator current to decay to zero. This is useful to evaluate the impact of the DC-offset on the generator current behaviour. The notable DC-offset in the transient, fifth-order model is seen in **Figure 4.6**. The presence of the DC-offset may lead to current magnitudes of around 6.0 p.u., whereas the current magnitude reaches only around 4.0 p.u. in the reduced, third-order model with a disregarded DC-offset.

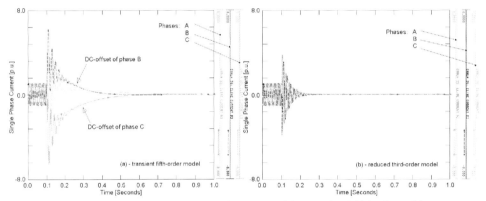

Figure 4.6 Single phase currents using: **(a)** - transient fifth-order model, **(b)** - reduced third-order model.

The DC-offset in the single phase currents lasts up to 400 ms for this generator. This characteristic time corresponds to complete de-excitation of the induction generator at the short-circuit fault (zero-voltage at the generator terminals). Note that the DC-offset is zero for the case of the reduced third-order model of induction generators. In this case, the predicted characteristic time of de-excitation is about 200 ms. In other words, the reduced, third-order model predicts faster de-excitation of the induction generator than will be reality in a physical induction generator. Prediction of faster de-excitation during the grid fault means prediction of larger demands for the reactive power to be absorbed by the induction generator when the fault is removed. The accurate prediction of the characteristic time of de-excitation during a grid fault is significant for reactive power dynamics and for the voltage profile of the induction generator.

The accurate prediction of generator current magnitude is a significant issue for generator protective relaying. When there is a risk of protective disconnection of the wind turbine due to excessive generator current (transients) then the reduced third-order model is not always accurate.

4.3.2.1   Balanced symmetrical events

In investigations of short-term voltage stability, the response of the power grid and the fault-ride-through capability of wind turbines are often examined with regard to 3-phase short-circuit faults in transmission power grids (Kundur, 1994; Bruntt et al., 1999). Such 3-phase short-circuit faults are balanced symmetrical events. Those are balanced as all three phases are subject to a short circuit at the same time and symmetrical because the fault occurs in all three phases of the symmetrical 3-phase system. The balanced symmetrical events are characterised by sudden and significant voltage changes which excite the fundamental-frequency transients (the DC-offset) in the generator current, see **Figures 4.5** and **4.6**.

This section explains that the transient, fifth-order model of induction generators is preferred to represent the response of induction generators during balanced symmetrical faults.

4.3.2.2   Unbalanced symmetrical events

An example of an unbalanced symmetrical events can be disconnection of a 3-phase line or removal of a 3-phase short-circuit fault. When a 3-phase line is disconnected, the single phases of the line are tripped one after another when the phase current crosses zero in each phase. Therefore, this is called an unbalanced event. This is a symmetrical event because all three phases are disconnected.

When the phase current in each single phase is disrupted at zero crossing, the DC-offset in the phase current is eliminated. Therefore the fundamental-frequency transients are not present in the stator current magnitude when the induction generator is subject to an unbalanced symmetrical fault (Pedersen et al., 2003).

Removing a short-circuit fault is often carried out by disconnection of the faulty line or the disconnection of lines connected to the faulty node. Therefore, the fundamental-frequency transients in the generator current are commonly suppressed when a 3-phase short-circuit fault is cleared. In **Figure 4.5**, there are no such excessive transients in the stator current when the 3-phase short-circuit fault is cleared.

When the power grid is subject to an unbalanced symmetrical fault, computations may be carried out with the use of the reduced, third-order model of induction generators (Pedersen et al., 2003). This model disregards the fundamental-frequency transients in the stator current. In this way, unbalanced events can be simulated using positive-sequence equivalents of power grids and the reduced, third-order model of induction generators.

When the positive-sequence equivalent is applied, the 3-phase representation of the power grid is replaced by a certain single-phase equivalent. Then, disconnection of a line in the positive-sequence equivalent of the power grid is computed modelling disconnection of all three phases at the same time. Since the reduced, third-order model disregards the fundamental-frequency transients in the stator current, the transients are eliminated and the correct result is gained. This technique is used in almost all existing power system stability tools, for example, PSS/E^{TM} and Powerfactory^{TM}.

4.3.2.3 Influence of saturation

Saturation represents non-linearity of the flux at excessive current through the winding. This phenomenon is relevant in investigations of harmonic emissions from the induction generator to the grid being subject to a fault resulting in a significant voltage drop at the generator terminals. This may also be relevant in investigations of the interaction between the generator rotor current and the rotor converter of the doubly-fed induction generators.

Ref. (Bolik, 2004) presents results for an induction generator modelled with representation of saturation and compares the results for the same induction generator modelled without saturation. **Figure 4.7** shows the no-load characteristic of the generator applied in Ref. (Bolik, 2004).

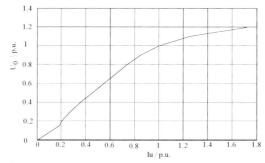

Figure 4.7   Measured no-load curve for modelling saturation of magnetizing inductance where U_0 is the no-load terminal voltage and I_U is no-load stator current, reprinted from Ref. (Bolik, 2004), Copyright (2004), with permission from Sigrid M. Bolik.

Figure 4.8 presents the terminal voltage of an induction generator, whilst **Figure 4.9** compares the results to the generator current as found in (Bolik, 2004). When a 3-phase short-circuit fault occurs and the terminal voltage drops, there may be a difference between the generator current computed with, and without, saturation. According Ref. (Bolik, 2004), this difference is measured to be 4% in the first current transient spike of 20 ms duration, when the voltage drops down to 0.2 p.u. As reported by (Sørensen et al., 2003), this difference in the first current transient spike of 20 ms can be up to 10% at a very efficient short-circuit fault. However, Ref. (Sørensen et al., 2003) does not specify the value of the voltage drop at the grid fault. When the current transients decay, the difference in the induction generator current modelled with, and without, saturation becomes small.

The results of Bolik (2004) and Sørensen et al. (2003) are important because they provide documentation that saturation effects in induction generators are not relevant in investigations of short-term voltage stability. The accuracy improvement due to the representation of saturation is below 10% for the first spike in generator current. This accuracy improvement in the computed current behaviour has only little influence on the positive-sequence voltage or the generator rotor speed of conventional induction generators, but requires larger computation time.

Besides, such accuracy improvement is seen when the voltage drops so much that the converter protection of the converter-controlled DFIG is activated before the current transients become as excessive as presented in (Bolik, 2004) and (Sørensen et al., 2003). Therefore, this will be inferior for investigations of short-term voltage stability, see also **Section 7.2.3**.

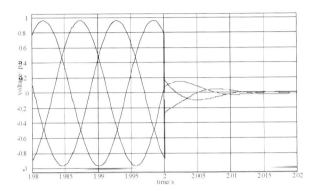

Figure 4.8   Terminal voltage of an induction generator, reprinted from Ref. (Bolik, 2004), Copyright (2004), with permission from Sigrid M. Bolik.

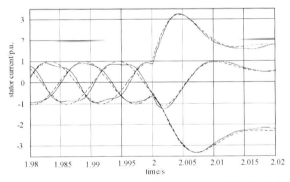

Figure 4.9   Stator current of an induction generator. Comparison between models with (solid line) and without saturation effects (dashed line), reprinted from Ref. (Bolik, 2004), Copyright (2004), with permission from Sigrid M. Bolik.

4.3.3   Significance of generator rotor inertia

In the case of induction generators, the voltage profile is coupled to the dynamic behaviour of the generator rotor speed during a short-circuit fault. The reduced, third-order model of induction generators may predict larger acceleration of such generators as the generator rotor speed starts to increase at the moment of the fault. In terms of the transient, fifth-order model, the generator rotor speed enters the short-circuit fault by a notch that leads to less acceleration of induction generators, which is caused by the braking torque of the transient stator flux of the induction generator, shown in **Figure 4.5**.

The dynamic behaviour of the generator rotor speed is given by the movement equation.

$$\frac{d\omega_G}{dt} = \frac{T_M - T_E}{H_G}. \tag{4.34}$$

So the impact of the braking torque results in the first transient spike in the electrical torque, T_E, on the generator rotor speed, ω_G, which depends on the generator rotor inertia constant, H_G. Subsequent acceleration of the generator rotor speed depends on the speed history (the notch value) at the fault entry. Ref. (Akhmatov and Knudsen, 1999) presents these results explaining the relation between the model details for induction generators, e.g. with or without representation of the stator flux transients, and the generator rotor inertia with regard to short-term voltage stability. Investigations described in (Akhmatov and Knudsen, 1999) were carried out on a simplified network equivalent representing the transmission power grid of Eastern Denmark. This simplified network equivalent contained ten nodes and four synchronous generators. The induction generator represented a large windfarm and was connected to the remote end of the system close to one of the four synchronous generators. The total power capacity of the large windfarm and the synchronous generator was 1400 MVA. The short circuit capacity of the transmission network in the connection point of the large windfarm was approximately 4000 MVA. The inertia constant of the induction generator rotor, H_G, was set to 0.5 s. According to data shown in **Table 4.3**, this corresponds to a typical rotor inertia constant of induction generators in the MW class. Investigations of (Akhmatov and Knudsen, 1999) showed notable discrepancy in the voltage behaviour between the transient fifth-order model and the reduced third-order model when the power capacity of the induction generator was increased (and the power capacity of the synchronous generator was simultaneously reduced). The results of the investigations are illustrated in **Figure 4.10**.

In most cases, the rotating system consists of more than just the generator rotor. The impact of braking torque on the notch value in the generator rotor speed occurring during the fault reduces as inertia increases. Therefore model-dependent differences are usually less when inertia becomes larger.

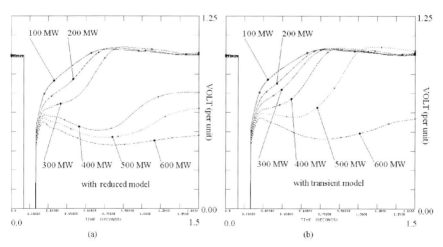

Figure 4.10 Comparison between model-dependent voltage curves in a light induction generator rotor: **(a)** – reduced third-order model, **(b)** – transient fifth-order model. Note that the reduced third-order model predicts voltage instability when the power capacity of the induction generator reaches 400 MW. When the transient fifth-order model is applied, voltage instability is predicted when the power capacity reaches 600 MW. Reprinted from (Akhmatov, 2003(b)), Copyright (2003), with permission from the copyright holder.

To demonstrate this, the above test was repeated in Ref. (Akhmatov and Knudsen, 1999) with increased values of H_G, thus decreasing the speed change caused by the initial transient torque. The rated power of the induction generator was kept at 500 MW. For comparison, a typical inertia constant of a wind turbine rotor, H_M, can be in the range of 2.5 s. When adding this to the generator rotor inertia constant, the total lumped inertia becomes 3 s. Results illustrating the influence of the total inertia on the voltage profile reached by using the transient, fifth-order model and using the reduced, third-order model are shown in **Figure 4.11**. The voltage curves start converging when the lumped inertia constant, H_G, reaches approximately 1 s and are almost fully converged when H_G is 3 s. This result means that in applications where the lumped inertia constant of the induction generator is relatively large, the reduced third-order model will produce almost correct results. Only when the generator rotor inertia constant becomes rather low (1 s or less), it is preferred to use the transient, fifth-order model in investigations of short-term voltage stability.

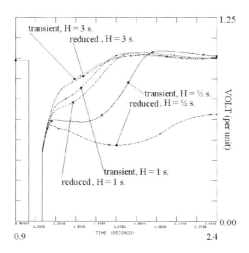

Figure 4.11   Model comparison - with varying inertia constants. Reported in (Akhmatov, 2003(b)), Copyright (2003), with permission from the copyright holder.

Results presented in this section are based on the assumption of the rotating masses consisting of one lumped mass. As explained in **Section 3.2.2**, the shaft systems of fixed-speed wind turbines provide relatively soft coupling between the rotor and the generator rotor. This "soft" coupling influences the dynamic behaviours of shaft torque as well as the generator rotor speed during grid disturbances.

4.3.4   Significance of shaft representation

Fixed-speed wind turbines shown in **Figure 4.12** are complex electromechanical constructions converting the mechanical power of the rotor into active power in the generator. Fixed-speed wind turbines are characterised by relatively soft coupling between light generator rotors and heavy turbine rotors. This coupling is arranged through wind turbine shaft systems. Accordingly to the description in **Section 3.2.2**, the shaft system is represented by a two-mass model. As the generator rotor is light, then it might still go through large speed variations during faulting events occurring in the power grid even though the wind turbine rotor is not accelerated much.

The acceleration of the generator rotor will be even larger because it will be accelerated not only by mechanical power from the wind turbine rotor, but also from tension in the shaft – the tension which is released when the electrical torque is either reduced or lost due to the short-circuit fault in the power grid. This topic is explained in detail in **Section 5.5**.

In order to get a full impression of the importance of the fundamental-frequency transients in the stator current and of the two-mass model of the shaft system, four comparative simulations were performed in (Akhmatov and Knudsen, 1999), applying the simplified network equivalent with four synchronous generators and a large windfarm. The fault event was a 100 ms short-circuit fault at the induction generator terminals. However, in this case the rating of the induction generator was 900 MW and the synchronous generator was 500 MVA. This change was made in order to get a situation where the differences in the simulation results between the various models were large. Four different models of induction generators of fixed-speed wind turbines were examined.

1) Reduced third-order model with the two-mass shaft model.
2) Transient fifth-order model with the two-mass shaft model.
3) Reduced third-order model with the lumped mass model.
4) Transient fifth-order model with the lumped mass model.

Simulation results are shown in **Figure 4.13**. Results for the lumped mass model are almost identical, independent from generator models (reduced or transient). This outcome is expected accordingly to the results described in **Section 4.3.3**.

Figure 4.12   1.3 MW Siemens fixed-speed, active-stall controlled wind turbines equipped with induction generators. Photo copyright Siemens. Reproduced with permission from Siemens WPG.

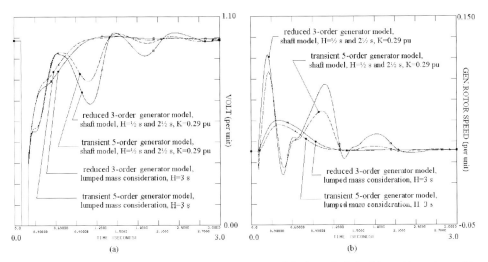

Figure 4.13   Comparison between models of induction generators of fixed-speed wind turbines: **(a)** – voltage, **(b)** – generator rotor speed. Reprinted from (Akhmatov, 2003(b)), Copyright (2003), with permission from the copyright holder.

Results for the two-mass model of the shaft system are however remarkable different when applying different generator models (reduced or transient). First, the curves produced with the use of the two-mass model differ from the curves obtained with the lumped mass model. Secondly, the two curves obtained with the two-mass model and different generator models are also notably different from each other unlike in the case of the lumped mass model. Notice that the worst case with regard to voltage stability is reached for the reduced, third-order generator model with the two-mass shaft model. This result shows that including the two-mass shaft model will reduce the stability margin in a more significant way than in the case of the lumped mass model. Notice also that the use of the transient fifth-order model together with the two-mass shaft model predicts improving the stability margin since the voltages are generally higher and less oscillatory than the voltages computed with the reduced third-order generator model with the two-mass shaft representation. The stability margin predicted with the use of the two-mass shaft model is still smaller than with the use of the lumped mass models, that is independent from what the type of generator model applied (reduced or transient).

Oscillations in the terminal voltage of the induction generator are caused by the mechanical oscillation of the small generator rotor against the larger inertia of the wind turbine rotor. The presence of such oscillations in the AC voltages means that the mechanical shaft oscillation may excite other modes of oscillation in the entire AC system; e.g. power oscillations which normally have low frequencies. Note that the torsion of the shaft system has significant impact on the voltage behaviour and on the voltage stability of fixed-speed wind turbines equipped with induction generators.

Representation of the shaft systems using the two-mass model reduces the stability margin, but this is partly counteracted by the transient, fifth-order model increasing the stability margin of induction generators. These results clearly show that the induction generators models of fixed-speed wind turbines must contain the two-mass representation of the shaft and the fundamental-frequency transients in the stator current.

4.3.5 Aerodynamic rotor model

Providing the complete dynamic model of fixed-speed wind turbines applied in investigations of short-term voltage stability requires also an aerodynamic rotor model and a model of the blade-angle control. This has been discussed in **Section 3.1**. The aerodynamic rotor model produces the coupling between the rotor speed, ω_M, and the mechanical torque of the rotor, T_M, applied to the shaft system and which accelerates the generator rotor.

4.3.6 Protective relay modelling

The protective relay system of a fixed-speed wind turbine monitors several electrical and mechanical parameters such as the terminal voltage, the stator current, the grid frequency and the (generator rotor) speed. The protective relay system orders disconnection of the wind turbines when at least one of the monitored parameters exceeds its relay settings. The data for the relay settings can be provided by the wind turbine manufacturer. **Table 5.2** in **Section 5.8** gives typical relay settings for fixed-speed wind turbines that can be applied in investigations of short-term voltage stability when the exact relay settings are unknown. Once given, the order to disconnect and stop the wind turbine cannot be cancelled. The wind turbines will be automatically reconnected and restarted 10 to 15 min. afterwards.

4.4 Model validation

Model validation is required and must demonstrate the accuracy and reliability of the developed model and explain possible discrepancies between the model and experimental results. Such measurements may be either from planned experiments carried out under controlled conditions (Pedersen et al., 2003) or from forced events so long as measurements are of sufficient resolution and accuracy (Akhmatov, 2005(b)).

Models of fixed-speed wind turbines equipped with induction generators must contain:

1) Transient fifth-order model of induction generators,
2) Two-mass model of the shaft system.

The primary target of the model validation must be evaluation of the generator model (reduced or transient) as well as of the shaft system representation (two-mass or lumped) as part of the dynamic model of fixed-speed wind turbines. The validation from planned experiments is to be preferred.

4.4.1 Case applied for validation

The results of the planned experiment carried out at the Nøjsomheds Odde windfarm, Eastern Denmark, were applied in the validation of the model of fixed-speed wind turbines equipped with induction generators (Raben et al., 2003). The Nøjsomheds Odde windfarm contains twenty-four Siemens fixed-speed, active-stall controlled wind turbines. The rated power of a single wind turbine is 1 MW and details are shown in **Table 4.4**. Part of the Nøjsomheds Odde windfarm is shown in **Figure 4.14**.

Parameters	Values
Rated power	1000 kW (strong wind) / 200 kW (light wind)
Synchronous rotor speed	22 RPM / 15 RPM
Rotor diameter	54.2 m
Gear-ratio	1:69
Rated voltage	690 V
Rated electric frequency	50 Hz

Table 4.4   Selected data for the Siemens 0.2/1 MW wind turbine, reprinted from Ref. (Raben et al., 2003), Copyright (2003), with permission from Multi-Science Publishing Company.

Figure 4.14   Part of the Nojsomheds Odde windfarm, Eastern Denmark, reprinted from Ref. (Raben et al., 2003), Copyright (2003), with permission from Multi-Science Publishing Company.

Such experiments are planned very carefully. Security of the personnel participating in the experiments is a key issue. The experiments must neither affect operation of the entire power grid nor cause damage to the equipment. Before the experiments are carried out, participants must receive permission from the relevant authorities, the local power company, the windfarm owner, and the wind turbine manufacturer (Raben et al., 2003). Obviously, the windfarm might not be subject to a 3-phase short-circuit fault, though the results of such an experiment would have a significant value for model validation. The wind turbines must also keep working during the whole experiment to provide a record of relevant measured parameters. Then, the experiment must not cause sub-sequent disconnection of the wind turbine.

The experimental work presented by (Raben et al, 2003) was undertaken in co-operation with the Danish power company SEAS-NVE who planned and carried out the experiment, the consulting company Hansen & Henneberg (Copenhagen, Denmark) who assisted with measurements, the manufacturer Siemens who contributed with wind turbine data and the power distribution company NESA who assisted with planning and simulations. As the experiment could not have a destructive character, the selected wind turbine in the Nøjsomheds Odde windfarm was subject to the tripping and reconnection sequence.

To avoid activation of the protective relays by excessive current transients at reconnection, the experiment is held in a light-wind situation. To approach faulted conditions which may cause high current transients, the duration of the tripped operation was set to 500 ms. Note that typical duration of a 3-phase short-circuit fault in the Danish transmission grid is below 100 ms.

The same kind of wind turbines from Siemens are commissioned at the Middelgrund offshore windfarm (twenty 2 MW wind turbines) and the Rødsand 1 / Nysted offshore windfarm (seventy-two 2.3 MW wind turbines). The Middelgrund offshore windfarm is shown in **Figure 5.11**. From this point of view, the presented validation has a practical relation to modelling of these large off-shore windfarms.

4.4.2   Experimental work

The Nøjsomheds Odde windfarm is separated into four sections which are arranged along 10 kV underground cable sections with six wind turbines in each section. Through the 0.7 / 10 kV transformers, the wind turbines are connected to the internal network of the windfarm. The windfarm is, then, connected to the local 50 kV distribution power system through a 10 / 50 kV transformer. Through a 50 / 132 kV transformer, the windfarm feeds into a transmission grid of the Danish island of Lolland shown in **Figure 4.15**. The short circuit capacity in the connection point to the transmission grid (132 kV voltage level) is 350 MVA.

A part of the windfarm network is shown in **Figure 4.15**. Before the experiment started, the induction generators and the no-load capacitors of wind turbines WT 02 to WT 06 were tripped and these wind turbines were temporarily stopped. When the experiment was completed, these wind turbines were reconnected to the grid and continued normal operation. Due to this procedure the induction motors of these temporarily disconnected wind turbines were still in operation providing forced cooling of the induction generators and keeping other vital functions of the temporarily disconnected wind turbines on stand-by. Only the wind turbine WT 01 in the windfarm section 1 was in operation during the experiment.

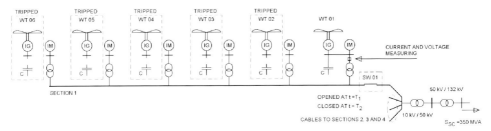

Figure 4.15   A fragment of a windfarm network with wind turbine WT 01 relating to a tripping – reconnection experiment, reprinted from Ref. (Raben et al., 2003), Copyright (2003), with permission from Multi-Science Publishing Company.

The tripping – reconnection experiment was carried out in a light wind to minimise the current transients at re-connection and, then, reach successful reconnection of the wind turbine WT 01 to the power grid. At the time of tripping, T_1, the switch SW 01 was opened and section 1 (with the wind turbine WT 01 in it) was in island operation with the induction motors and internal network of the section given by the impedance of the cables and transformers, and the cable charging.

At reconnection time, T_2, the switch SW 01 was closed and section 1 was reconnected to the power grid. In the experiment, the voltage and the current at the low-voltage side of the 0.7 / 10 kV transformer of wind turbine WT 01 were measured, as marked in **Figure 4.15**. The sampling frequency of the measurements was 1 kHz, ensuring that any delay in the measured signals produced by the equipment could be disregarded.

In the island operation, the no-load capacitor of the wind turbine generator WT 01 was kept connected to the grid, so as to reduce any possible voltage drop at the terminals of the wind turbine (WT 01). The measured behaviour of the phase current, I_L, and the phase-to-phase voltage, U_{LL}, are shown in **Figure 4.16**. In island operation, the current of the no-load compensated induction generator of wind turbine WT 01 was not zero because the induction motors in section 1 absorbed some of the active power produced by the wind turbine generator. An amount of the reactive power was also exchanged through the internal network of the section, taking into account the no-load impedance of the transformers and cable charging.

In island operation and shortly after reconnection, higher harmonics of the fundamental frequency were observed in the measured current. Such higher harmonics were caused by the induction machines, according to the explanation given by (Pedersen et all, 2000). The measured voltage showed no (significant) drop in island operation.

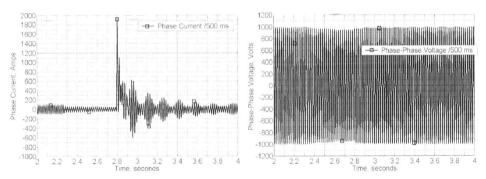

Figure 4.16 Measured phase-current and phase-to-phase voltage. Time of island operation is 500 ms. Reprinted from Ref. (Akhmatov, 2003(b)), Copyright (2003), with permission from the copyright holder.

At the time of tripping, $t = T_1$, there was no dc-offset in the measured phase current. However there was notable dc-offset in the measured current at the time of reconnection, $t = T_2$. The dc-offset at tripping was suppressed because opening of switch SW 01 occurred separately in the three phases so that the single phases were disconnected at the respective moments when the phase-currents crossed zero in each single phase. This was an unbalanced symmetrical event, see **Section 4.3.2.2**. Closing of switch SW 01 occurred abruptly at the same time, $t = T_2$, in all the three phases. This abrupt switching excited the notable dc-offset in the measured phase current at the moment of reconnection. This was a balanced symmetrical event, see **Section 4.3.2.1**.

In island operational mode and after reconnection, the voltage magnitude was almost unchanged. The machine current showed fluctuating behaviour after the wind turbine was reconnected to the power grid. The natural frequency of the current fluctuations was about 7 Hz. These fluctuations cannot be explained only by the dynamic behaviour of the induction generator itself. The value of the natural frequency indicates that the current fluctuations could relate to the torsion oscillations of the shaft in wind turbine WT 01.

This experimental case is interesting because it highlights an event which excites the transient behaviour of the generator current, as well as an event suppressing the current transients. Combined with the current fluctuations observed, this case is very relevant for validation of wind turbine models.

4.4.3 Choosing a simulation tool

Computations were carried out with the use of the simulation tool PSS/E^{TM}. At the moment of investigations, the simulation tool PSS/E^{TM} did not contain any detailed modelling of fixed-speed wind turbines. Therefore the user-written model of fixed-speed wind turbines implemented in the tool PSS/E^{TM} as presented in (Akhmatov and Knudsen, 1999) was used in validation work.

The simulation tool PSS/E^{TM} is commonly applied in investigations of power system stability of large power systems. The complexity of the investigated power system can be almost unlimited, e.g. with almost unlimited number of nodes, lines and generators. Power system models in PSS/E^{TM} are represented with the use of positive-sequence equivalents. This simplification is reasonable and acceptable in investigations of power system stability though this may introduce restrictions on representation of the transient behaviour in electric machines and, during transient events.

As the positive-sequence equivalents of power systems are applied, the simulations are restricted basically to the examination of symmetrical, 3-phase faults and events. PSS/E^{TM} is the fundamental-frequency tool, which implies that frequency deviations occurring during grid events must be in the range of to 10% of the nominal grid frequency. Higher harmonics of the fundamental frequency are not represented in simulations. Furthermore, the standardised generator models within PSS/E^{TM} are reduced-order models where the fundamental-frequency transients in the machine current are disregarded. This is in agreement with the idea of such stability investigations of large power systems (Kundur, 1994).

4.4.4 Computational work

The simulation work was started by creating a grid model consisting of section 1 with a grid connected wind turbine WT01, the induction motors of the disconnected wind turbines WT02 to WT06, and the cables and the transformers connecting the machines to the transmission grid equivalent. All relevant data for the network, the wind turbines and the motors were either received from the power grid company and the wind turbine manufacturer, or estimated using data for similar electric devices (Raben et al., 2003). The load-flow solution found for the grid equivalent was in agreement with measured load-flow of the power grid applied in the validation work. An accurate load-flow solution was important to ensure the start of the dynamic simulation was from the right steady-state condition.

In post- dynamic simulations, the no-load capacitor of wind turbine WT01, the transformers and cables were all represented as quasi-dynamic models of the tool PSS/E^{TM}. Induction motors were modelled using the reduced, third-order model CIMTR4, that is a standardised part of PSS/E^{TM}.

In the validation work, the following models of fixed-speed wind turbines were compared.

1) Transient, fifth-order model of the induction generator and the two-mass model of the shaft system.
2) Transient, fifth-order model of the induction generator and the lumped-mass model of the rotating parts.
3) Reduced, third-order model of the induction generator and the two-mass model of the shaft system.

Figure 4.17 presents the current-phasor magnitude and voltage-phasor magnitude at the low-voltage side of the wind turbine transformer computed using the transient, fifth-order model of the induction generator and the two-mass model of the shaft system. Additionally, the generator rotor speed is shown in **Figure 4.17.c**. The simulated curves of this model are in agreement with the measured data shown in **Figure 4.16**. The dc-offset is present in the simulated current-phasor magnitude, which is similar to the behaviour of the measured phase-current. At the moment of re-connection, the simulated current-phasor magnitude reaches 14 times that of the current magnitude in undisturbed operational conditions. This simulated result is close to the ratio reached by the measured phase-current at the moment of re-connection (approximately 18 times bigger).

The discrepancy between the measured and the simulated results can be due to several reasons such as:

1) The fundamental-frequency current transients in the no-load capacitor were disregarded in the simulations, but such transients were present in reality.
2) The phase-current was measured in a single phase whereas the computed current was the current-phasor, e.g. an effective 3-phased value. Therefore these values were close to each other, but not necessarily matched exactly.

The simulated and measured behaviour of the current showed the same fluctuating behaviour with a natural frequency of approximately 7 Hz. The fluctuations seen in the current behaviour related to shaft torsion, seen from the generator rotor speed and plotted in **Figure 4.17.c**.

The measured and simulated oscillating current behaviour was characterised by different damping. This difference is explained by the damping coefficients, D_M and D_G, being set to zero in simulations, but damping and friction were present in the rotating parts of the wind turbine construction. However, this discrepancy is deemed negligible. Thus, the assumption that the damping coefficients in simulations were zero seems to be reasonable.

The lumped-mass model corresponds to the assumption that the shaft system of the wind turbine is infinitely stiff, i.e. $K_S \rightarrow +\infty$, and that the rotating parts of the wind turbine construction are lumped together. According to (Akhmatov et al., 2000(c)), this corresponds to $H_G + H_M$. The simulated behaviour of the current-phasor magnitude and the generator rotor speed is shown in **Figure 4.18**. This simulated behaviour is not in agreement with the real situation as the lumped-mass model was applied to represent the rotating parts of the wind turbine. When the lumped-mass model was applied, the following discrepancies were seen.

1) Fluctuations in the simulated behaviour of the current-phasor magnitude and the generator rotor speed were not seen when reconnection occurred.

2) The magnitude of the current-phasor overshoot at the moment of re-connection was much lower (approximate ratio of 11 compared to the undisturbed current magnitude) than predicted by the two-mass model and seen in the experiment. This model-dependent discrepancy of the current can be important for the relay modelling of wind turbine generators.

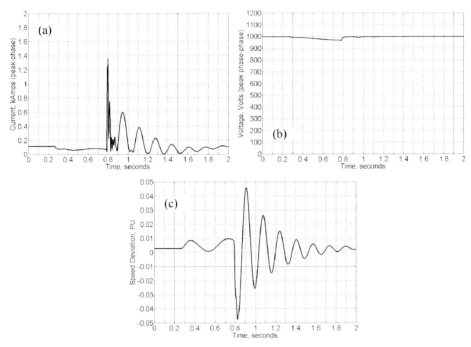

Figure 4.17 Simulated behaviour of: (a) - current and (b) - voltage magnitude at "the point of measurement" and (c) - simulated generator rotor speed deviation (minus slip), using the transient, fifth-order model of induction generator and the two-mass model of shafts. Reprinted from Ref. (Akhmatov, 2003(b)), Copyright (2003), with permission from the copyright holder.

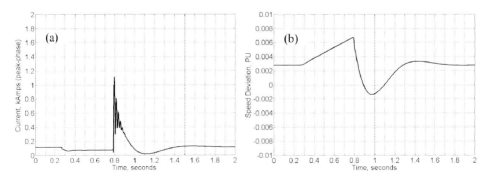

Figure 4.18 Simulated behaviour of: **(a)** - current-phasor magnitude and **(b)** - generator rotor speed deviation (minus slip), using the transient, fifth-order model of induction generators and the lumped-mass model of rotating parts. Reprinted from Ref. (Akhmatov, 2003(b)), Copyright (2003), with permission from the copyright holder.

Figure 4.19 shows the current-phasor magnitude computed with the use of the reduced third-order model of induction generators and using the two-mass model of shafts. As expected, a discrepancy was seen in the machine current behaviour because the fundamental-frequency transients in the machine current were disregarded in this model. As there was no dc-offset in the simulated behaviour of the current-phasor magnitude, the overshoot at the moment of re-connection was only 8 times higher when compared to undisturbed operation. This ratio is much lower than predicted with the use of the transient fifth-order model of induction generators or seen in the experiment.

Fluctuations in the computed current-phasor magnitude were present, characterised by the natural frequency of approximately 7 Hz. Such fluctuations were also seen in the measured current when reconnection occurred and were caused by shaft torsion oscillations.

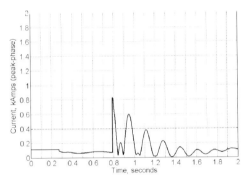

Figure 4.19 Simulated current-phasor magnitude using the reduced, third-order model of induction generators and the two-mass model of shafts. Reprinted from Ref. (Akhmatov, 2003(b)), Copyright (2003), with permission from the copyright holder.

4.4.5 Summary of validation work

The dynamic model of fixed-speed wind turbines equipped with induction generators, implemented using the simulation tool PSS/E^{TM} as the user-written model, was validated from the measurements of the tripping - reconnection experiment. Though this experiment was not characterised by the conditions of a short-circuit fault with significant voltage drop, it was useful and relevant for the validation of a simulated response reached with the use of the models of different complexities. In the experiment, a selected wind turbine in the Nøjsomheds Odde windfarm was brought into island operation within a section of the internal network of the windfarm. Shortly after tripping, the windfarm was reconnected to the entire power system. The experiment was carried out on the same type of wind turbines erected at the Rødsand 1 / Nysted offshore windfarm in Eastern Denmark, but of a lower rated power than those in the offshore windfarm. The main target of this work was the full-scale validation of the dynamic wind turbine model. Therefore wind turbine models of different complexities were examined and compared to actual measurements. Models of different complexities could be arranged according to the accuracy of their output, i.e.:

1) The dynamic wind turbine model containing the transient fifth-order model of induction generators and the two-mass model of the shaft system - being the most accurate.

2) The model containing the reduced third-order model of induction generators and the two-mass model of the shaft system.

3) The model containing the transient fifth-order model of induction generators and the lumped-mass model of the rotating system.

This result shows that the two-mass model of shaft systems is essential to reach an accurate representation of fixed-speed wind turbines in investigations of short-term voltage stability. This result also shows the relevance of the transient fifth-order model of induction generators applied to investigations of wind turbine responses to grid disturbances. Only small discrepancies between measured data and the simulated results were present. Such discrepancies were explained in (Raben et al., 2003) by uncertainties in the wind turbine data (the data for damping are not given) and restrictions of the simulation tool PSS/E^{TM}.

The dynamic wind turbine model containing the transient fifth-order model of induction generators and the two-mass model of shaft systems was used for the grid planning work for connection of the Rødsand 1 / Nysted offshore windfarm to the eastern Danish power grid (Akhmatov et al., 2003(a)).

4.5 Summary

Modern power systems in industrialised countries are 3-phase AC- systems and modern fixed-speed wind turbines equipped with conventional induction generators are connected to such 3-phase AC- systems. In the 3-phase AC- systems, the voltage and current in each phase oscillate with a grid frequency that is either 50 Hz (in Europe and Australia) or 60 Hz (in USA). Induction generators convert the mechanical power produced by the wind turbine rotor into active electrical power supplied to the grid. Though the active power of each single phase oscillates with the double of the grid frequency, the total active power of the 3-phase symmetrical system is constant at constant magnitudes of stator voltage and current. The mechanical power of the rotor can, then, be continuously balanced by the total active power of the 3-phase induction generator without detriment to the rotor speed.

Conventional induction generators with a short-circuited rotor circuit supply active power to the grid and absorb reactive power from the grid. This implies that the conventional induction generators are excited from the power grid. Such induction generators cannot control their excitation and therefore require that the terminal voltage be close to their rated voltage. Shunt capacitors are often applied at the induction generator terminals to compensate for the reactive power absorption of the induction generators. In this way, the reactive power absorption from the grid itself is reduced and the power factor of the fixed-speed wind turbines is improved.

In conventional induction generators with a short-circuited rotor circuit the active and reactive power are strongly coupled to the slip being the relative difference between the electrical and the mechanical frequencies. In normal operation of the power grid, the slip of conventional induction generators may be up to about 2%. Then the rotor speed of the wind turbines equipped with such conventional induction generators is almost fixed to the electrical synchronous speed of the power grid which is why such wind turbines are denoted fixed-speed.

The model of induction generators is part of the dynamic wind turbine model applied in investigations of short-term voltage stability. The 3-phase model of induction generators is difficult to use in such stability investigations that may involve large power systems with a large number of generators. In steady-state operations, the states and the state derivatives of the 3-phase model oscillate with the grid frequency and the state derivatives are not zero. Instead it is convenient to apply a model of induction generators in the synchronously rotating reference frame. When transformed to the synchronously rotating reference frame, the states (i.e. the fluxes) of induction generators are constant in steady-state. The state derivatives of the model are, then, zero in steady-state which provides easy and obvious initialisation of the model states. There are also other basic advantages to apply the models in the synchronously rotating reference frame which are more described in detail in (Kundur, 1994)

In short-term voltage stability investigations, focuses are put onto the fundamental (positive-sequence) component of the voltages through large power systems with an arbitrary number of generators. To reduce the computation time and adapt the generator models to positive-sequence equivalents of the power system investigated, generator models are usually represented with their reduced non-transient models. In the case of induction generators, investigations of short-term voltage stability are usually carried out using the reduced, third-order model. This model disregards fundamental-frequency transients in the stator flux and, then, in the rotor current. This simplification may however lead to prediction of more overspeeding of the generator rotor at voltage drops in the power grid. Due to strong coupling between the electrical and mechanical parameters in induction generators, more overspeeding results in prediction of larger reactive power demands and more difficult voltage recovery than in physical induction generators.

Disregarding the fundamental-frequency transients in the stator current may also result in inaccurate predictions of stator current magnitude at grid disturbances. This may lead to inaccurate response of the relay models of wind turbines as the relaying can be sensitive to such excessive current transients.

In this presentation, induction generators are modelled using the transient, fifth-order model. This model respects the fundamental-frequency transients in the stator flux and current and accurately predicts behaviour of the generator speed in short-circuit faults. Such fundamental-frequency models can be adapted to work together with positive-sequence equivalents of the power grids without notable increase of the computation time (Akhmatov and Knudsen, 1999).

The largest increase in accuracy is gained when replacing the reduced, third-order model by the transient, fifth-order model of induction generators. This accuracy increase shows up in all parameters of induction generators such as the generator rotor speed, terminal voltage and machine current. When comparing the machine current magnitude in a significant voltage drop at the generator terminals, the difference in the results between the reduced, third-order model and the transient, fifth-order model may be in the range 30 to 40% (the voltage drops to 0.2 p.u.). The accuracy can further be improved by representation of saturation effects, but the accuracy improvement in the current behaviour is up to a few percent only at a voltage drop to 0.2 p.u. (Bolik, 2004).

When the inertia of the generator rotor itself is large, the difference between the results predicted using the reduced, third-order model and the transient, fifth-order model is reduced significantly (except for the current behaviour). The model-related difference is caused by the notch in the generator rotor speed at the time of entry of the short-circuit fault. As the inertia increases, the notch is reduced that eliminates the model-related difference with regard to the speed and the voltage of induction generators.

The inertia increase would be present when a light induction generator is driven by a heavy turbine connected directly to the generator rotor shaft. However the heavy rotor of the wind turbine is connected to the light generator rotor via the shaft system that provides a relatively soft coupling between the rotor and the generator rotor. Evaluation of the induction generator models as part of the dynamic model of fixed-speed wind turbines is therefore made in the following cases.

1) Reduced third-order model with the two-mass shaft model.
2) Transient fifth-order model with the two-mass shaft model.
3) Reduced third-order model with the lumped mass model.
4) Transient fifth-order model with the lumped mass model.

The models were evaluated using measurements and results of already validated models. The model consisting of the transient, fifth-order model of induction generators and the two-mass model of shafts was found to be in good agreement with the measurements available from experiments, for example, carried out at the Danish windfarm at Nøjsomheds Odde (Akhmatov, 2003(b)). The other three models gave results with less accuracy.

5 Voltage stability of fixed-speed wind turbines

The electrical and mechanical parameters and controllability of fixed-speed wind turbines equipped with conventional induction generators influence the short-term voltage stability and ride-through capability. This can be explained in the terms of *the dynamic stability limit* of grid-connected induction generators. This important relation is discussed in Ref. (Akhmatov et al., 2000(c)) and confirmed by investigations of (Holdsworth et al., 2001; Holdsworth et al., 2004).

5.1 Electric torque versus speed characteristic

The static characteristic of electric torque versus generator rotor speed of induction generators is given by the relation (Kundur, 1994).

$$T_E(\omega_G) = \frac{U_S^2}{\omega_G} \cdot \frac{R_T(\omega_G)}{R_T^2(\omega_G) + X_T^2(\omega_G)} . \tag{5.1}$$

The machine impedance viewed from the terminals of the induction generator is:

$$Z_T = R_T + jX_T \ ,$$

$$R_T(\omega_G) = R_S + \frac{\dfrac{R_R}{\omega_G - 1} \cdot X_M^2}{\left(\dfrac{R_R}{\omega_G - 1}\right)^2 + (X_M + X_R)^2} \ , \tag{5.2}$$

$$X_T(\omega_G) = X_S + \frac{X_M \cdot \left(\left(\dfrac{R_R}{\omega_G - 1}\right)^2 + X_R \cdot (X_M + X_R)\right)}{\left(\dfrac{R_R}{\omega_G - 1}\right)^2 + (X_M + X_R)^2} .$$

Here R_S and R_R are the resistance of the stator and rotor circuits, respectively, and X_S, X_M and X_R are the reactance of the stator, magnetising and rotor circuits, respectively, and all the values are in the stator quantities. The value of $(1 - \omega_G)$ gives the generator rotor slip, which is negative for generator-operation mode. The generator rotor speed is in p.u.

The electric torque, T_E, is a function of the terminal voltage. A number of electric torque versus speed curves are shown in **Figure 5.1**. These curves are plotted for different fixed terminal voltages, U_S, and there is no dependence between the terminal voltage, U_S, and the generator rotor speed, ω_G. The electric torque versus speed curve for a given induction generator can be gained from the manufacturer. Such a curve is typically plotted for the rated terminal voltage $U_S = 1.00$ p.u., independently from the reactive absorption of the induction generator (at excessive generator rotor speed). This may correspond to operation of the given induction generator in the grid with an infinite short-circuit capacity, $S_{SC} \to \infty$.

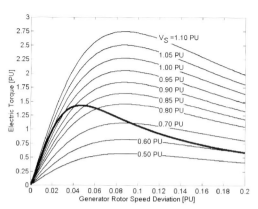

Figure 5.1   The electric torque versus generator rotor speed curves computed at a constant terminal voltage (thin solid line) and when the grid is with a finite short-circuit capacity (**thick solid line**). Reprinted from (Akhmatov, 2003(b)), Copyright (2003), with permission from the copyright holder.

The short-circuit capacity of physical power grids is always finite. When an induction generator is connected to the grid with a finite short-circuit capacity, the electric torque versus speed curves are influenced by a number of factors with relation to the power grid (Akhmatov et al., 2000(c)). The initial value of the terminal voltage is normally around 1.00 p.u., but not necessarily exactly 1.00 p.u. In the grid with a finite short-circuit capacity, the terminal voltage becomes dependent on the generator rotor speed because the reactive absorption of the induction generator, Q_E, is speed-dependent.

$$Q_E = U_S^2 \cdot \frac{X_T(\omega_G)}{R_T^2(\omega_G) + X_T^2(\omega_G)}.$$ (5.3)

The reactive power supplied to the terminals by the compensating capacitor is defined as:

$$Q_C = X_C \cdot U_S^2,$$ (5.4)

with the impedance of the compensating capacitor X_C. When the generator rotor speed increases, the reactive absorption of the induction generator also increases, and the terminal voltage decays, according to the relation:

$$\frac{\Delta U_S}{U_S} \sim \frac{-\Delta Q_E + \Delta Q_C}{S_{SC}}.$$ (5.5)

Here $(-\Delta Q_E + \Delta Q_C)$ denotes variations in the reactive power flow towards the connection node of the induction generator, and ΔU_S is the corresponding voltage variation. The changes in the reactive power of the induction generator are speed-dependent, $\Delta Q_E(\omega_G)$. Therefore the dependence of the terminal voltage on the generator rotor speed, $U_S(\omega_G)$, must be taken into account when computing the electric torque versus speed curves of the induction generator connected to the grid with a finite short-circuit capacity. In **Figure 5.1**, note the difference between the curve plotted for a grid with a finite short-circuit capacity and the curve plotted on the assumption of $U_S = 1.00$ p.u.

1) Since the power grid is with a finite short-circuit capacity, the kip-torque, e.g. the maximum torque, and the kip-speed are reduced when compared to the respective values gained from the curve plotted at the fixed terminal voltage $U_S = 1.00$ p.u.
2) When the generator rotor speed exceeds the kip-speed, the electric torque decays significantly faster than for the curve of the electric torque plotted at the fixed terminal voltage $U_S = 1.00$ p.u.
3) When the short-circuit capacity of the grid is reduced, the power grid is weak. In the weak power grid, the kip-speed and the kip-torque are reduced when compared to the case of the strong grid.
4) In the weak power grid, the electric torque decays faster at increasing generator speed above the kip-speed.

Note that electric torque versus speed curves of the same induction generator will show different behaviour when the generator is connected to grids with different short-circuit capacities. The different behaviour of the curves means gaining different values of the kip-speed, the kip-torque and the decay rate of the electric torque when the generator rotor speed exceeds the kip-speed.

5.2 Static stability limit of induction generators

The electric torque versus speed characteristic of a grid-connected induction generator is plotted in **Figure 5.2(a)**. At this stage, only the static stability of the induction generator is shown, whereas dynamic phenomena in the generator and shaft torsion are omitted.

The mechanical torque, T_M, is accelerating and the electrical torque, T_E, is decelerating. The electrical torque versus speed characteristic is given in Eq.(5.1) and the mechanical torque is defined by the relation:

$$T_M = \frac{P_M}{\omega_G}. \tag{5.6}$$

Figure 5.2(a) shows the mechanical torque versus speed curves for three operational situations which are (i) the rated operation level (100%), (ii) 75% of the rated operation and (iii) 50% of the rated operation. Each operational situation defines two different operational points of the induction generator with respect to stability. When the induction generator is, for instance, at the rated operation level, the steady-state operation is only possible in the two operational points marked in **Figure 5.2(a)** as (1) and (2). In these two operational points, the accelerating toque, T_M, and the decelerating torque, T_E, are equal. When considering which torque is accelerating or decelerating, only the operational point (1) is stable. As can be seen, all the possible steady-state operational points must be in the range from the synchronous speed, e.g. with zero speed deviation, to the kip-speed, ω_K.

When the induction generator operates at 50% of the rated power, similar considerations with regard to stability can be made. The operational points, where the decelerating, T_E, and the accelerating torque, T_M, are equal, are marked as (3), when the speed is below the kip-speed, respectively, as

(4), when the speed is above the kip-speed. All the possible steady-state operational points are again in the speed range below the kip-speed, ω_K.

In other words, the kip-speed, ω_K, defines the *static stability limit* of induction generators independently from the initial operational point of the generator. The kip- operational point is marked in the electric torque versus speed curve in **Figure 5.2(a)** as (K). When the generator rotor speed exceeds the kip-speed, this should result in excessive over-speeding of the induction generator which leads to voltage instability.

5.3 Dynamic stability limit of induction generators

Consider that an induction generator accelerates due to a grid fault. The fault can be a short-circuit fault that hinders the generator to feed the electric power, P_E, into the grid. Acceleration of the induction generator is described by the movement equation.

$$2 \cdot H_G \cdot \frac{d\omega_G}{dt} = T_M - T_E \quad . \tag{5.7}$$

When the grid fault is cleared, the induction generator will only be able to return to its regular operational point if the generator rotor speed, ω_G, does not exceed the so-called critical value of speed, ω_{CR}. From Eq.(5.7), the critical speed, ω_{CR}, is given by the relation.

$$T_M = T_E \implies \omega_{CR} \quad , \qquad \omega_{CR} \geq \omega_K \quad . \tag{5.8}$$

The critical speed, ω_{CR}, is defined by the cross-point between the curves of the decelerating torque, $T_E(\omega_G)$, and the accelerating torque, $T_M(\omega_G)$. The induction generator, when over-sped during the grid fault, will return to its regular operational point so long as the decelerating torque, T_E, exceeds the accelerating torque, T_M. If the critical speed, ω_{CR}, is exceeded, the induction generator continues to accelerate out of control. In this case, disconnection and use of an emergency brake will be necessary to stop the generator.

From **Figure 5.2(a)**, an induction generator can be accelerated above the kip-speed without reaching an excessive over-speeding point. When the induction generator is at its rated operation before the fault occurs, the critical speed is given by operational point (2). When the induction generator operates at 50% of the rated power before the fault, the critical speed is given by operational point (4). In other words, the critical speed of the induction generator depends of its operational point before the grid fault. When the actual operational point is below the rated operational point, the induction generator will show more stable behaviour during grid faults because the generator may accept more over-speeding and still remain stable. This is illustrated by simulations presented in **Figure 5.2(b)-(d)**. The induction generator maintains stable operation so long as the critical speed, dependent on the initial operational point, is not exceeded.

Note that this result cannot be explained in terms of the static stability limit definition. The static stability of induction generators in terms of exceeding the kip-speed, ω_K, is insufficient to understand dynamic behaviour of the power grids with a large amount of power supply from the induction generators.

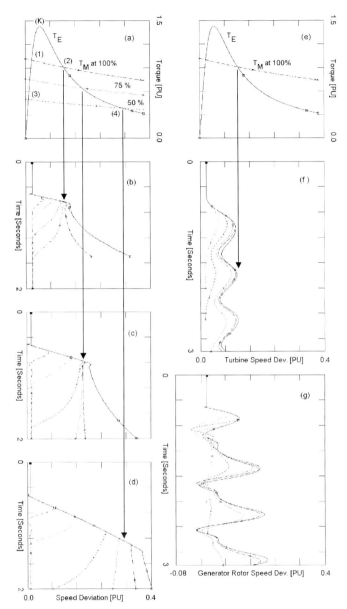

Figure 5.2   Definition of dynamic stability limit: **(a)** – curves of electrical and mechanical torque versus speed of an induction generator at different operational points, **(b)** – generator rotor speed of an induction generator at rated operation with P_E =2 MW and H_G =0.5 s subject to a grid fault, **(c)** – as (b), but at 75 % of rated operation, **(d)** – as (b), but at 50 % of rated operation. Dynamic stability limit in the case of wind turbines: **(e)** – curves of electrical and mechanical torque versus speed at rated operation, **(f)** – rotor speed at the grid fault with P_E =2 MW, H_G =0.5 s, H_M =2.5 s, K_S =0.3 PU/el.rad., **(g)** – generator rotor speed. Reprinted from Ref. (Akhmatov, 2003(b)), Copyright (2003), with permission from the copyright holder.

5.3.1 Critical speed as dynamic stability limit

The use of the critical speed, ω_{CR}, as a measure of dynamic stability of induction generators is introduced in Ref. (Akhmatov et al., 2000(c)). Ref. (Akhmatov et al., 2000(c)) presents an example of a power grid with two windfarms. The diagrammatic sketch of the power system is shown in **Figure 5.3**. This is a simplified model of the distribution power grid on the Danish island of Lolland in 1998. In this simulation example, the induction generators of the two windfarms are set to 75% of their rated power. The short-circuit capacity at the terminals of the two windfarms is different due to different electrical impedances in the lines and cables applied in the distribution power grid. This implies that the induction generators of the two windfarms have different dynamic stability limits as shown in **Figure 5.4**. The critical speed of the windfarms is about 7% over the synchronous speed (the generator rotor slip is -7%).

The duration of a short-circuit fault in the 132 kV transmission grid is chosen to be so long that the induction generators of windfarm 1 get exceeded the critical speed, whereas the induction generators of windfarm 2 do not. When the short-circuit fault is removed, the induction generators of windfarm 1 are excessively over-sped. The decelerating electrical torque is therefore insufficient to prevent more acceleration of the generators. The voltage in the vicinity of windfarm 1 decays, finally leading to voltage instability and disconnection of the windfarm.

The induction generators of windfarm 2 have not exceeded the critical speed. When the short-circuit fault is cleared, the generator rotor speed is, then, reduced and the voltage in the vicinity of windfarm 2 recovers. The simulated behaviour of the generator rotor speed and the terminal voltage of windfarms 1 and 2 are plotted in **Figure 5.5**. This example demonstrates how the dynamic stability limit works.

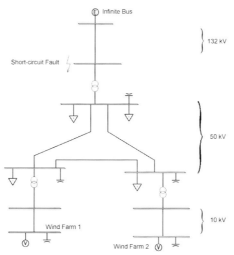

Figure 5.3   Diagram of a power distribution grid with two windfarms.

Definition of the dynamic stability limit is relevant when the arrangements and the control to improve the ride-through capability of fixed-speed wind turbines are to be investigated.

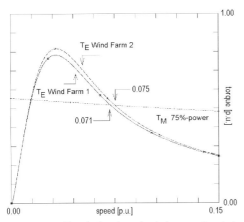

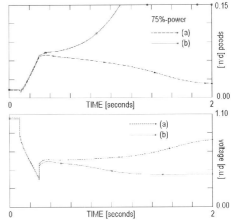

Figure 5.4   Electrical and mechanical torque in wind-farms 1 and 2. The marks give the critical speed in p.u. with regard to the synchronous speed.

Figure 5.5   Simulated generator rotor speed and terminal voltage of: **(a)** - windfarm 1 and **(b)** - windfarm 2.

5.3.2   Dependence on generator parameters

Some parameters of the induction generator influence the shape of the electrical torque versus speed curve, including the stator resistance, R_S, the stator reactance, X_S, the rotor resistance, R_R, the rotor reactance, X_R, and the magnetising reactance, X_M. Therefore these generator parameters will influence the dynamic stability of grid connected induction generators. Ref. (Akhmatov et al., 2000(c)) presents results of such a parameter study with regard to short-term voltage stability.

Figure 5.6 presents simulated curves of voltage, U_S, and generator rotor speed, ω_G, at different generator parameters, when the generator is subject to a 3-phase short-circuit fault of 100 ms duration. The investigations of (Akhmatov et al., 2000(c)) have shown that the critical speed limit is expanded and the dynamic stability of grid connected induction generators is improved when:

1) The rotor resistance, R_R, is increased.
2) The stator resistance, R_S, the stator reactance, X_S, the rotor reactance, X_R, and the magnetising reactance, X_M, are reduced.

5.4   Dynamic stability limit of wind turbines

The definition of dynamic stability limit will now be adapted to electricity-producing wind turbines that are two-speed systems. For this purpose, the lumped-mass model is applied to represent the movement of two-speed systems, for example the wind turbine shaft. The lumped-mass equivalent is seen as a mechanical system following the movement of a two-speed system as an integrated whole system (Akhmatov et al., 2000(a)). The movement equation of the lumped-mass equivalent is derived from the state equations of the two-mass model Eq.(3.46).

$$\begin{cases} 2\cdot(H_M+H_G)\cdot\dfrac{d\omega_L}{dt}=T_M-T_E \ , \\[2mm] \omega_L = \dfrac{H_G\cdot\omega_G+H_M\cdot\omega_M}{H_G+H_M} \ , \end{cases} \tag{5.9}$$

where the lumped inertia constant is (H_M+H_G) and the speed of the lumped-mass equivalent is ω_L.

Often, the rotor inertia constant, H_M, is much larger than the inertia constant of the generator rotor, H_G. On the assumption of $H_M \gg H_G$, the lumped-equivalent speed given by Eq.(5.9) can be reduced to the rotor speed.

$$\omega_L \approx \omega_M \ . \tag{5.10}$$

This implies that the critical speed, ω_{CR}, which defines the dynamic stability limit of the grid-connected wind turbines, must relate to the rotor speed, ω_M, rather than to the generator rotor speed, ω_G. The simulation results shown in **Figure 5.2(e)-(g)** confirm this adaption made for the case of electricity-producing wind turbines. As shown, the generator rotor speed, ω_G, may exceed the critical speed, ω_{CR}, without any consequences for excessive over-speeding of the wind turbines. The wind turbines become excessively over-sped when the rotor speed, ω_M, approaches the critical speed, ω_{CR}.

5.5 Shaft relaxation process

Figure 5.7 illustrates shaft system operation during a grid fault. In normal operation of the power grid, the soft shafts of the electricity-producing wind turbines are twisted through twist angle θ_S. When the grid is subject to a short-circuit fault, the grid voltage, U_S, and then the electrical torque, T_E, are reduced. The shaft, then, starts to relax. As explained in Ref. (Akhmatov et al., 2000(a)), this shaft relaxation contributes to more acceleration of the generator rotor during the grid fault.

The potential energy accumulated by the twisted shaft, with stiffness K_S is:

$$W_S = \tfrac{1}{2}\cdot K_S\cdot\theta_S^2 = ? \ \cdot\dfrac{T_M^2}{K_S} \ , \tag{5.11}$$

considering θ_S to be the initial twist of the shaft and neglecting damping. In normal operation of the grid, the kinetic energy of the rotating two-mass system is given by the relation.

$$E = H_M\cdot\omega_M^2 + H_G\cdot\omega_G^2 \ , \tag{5.12}$$

where ω_M and ω_G denote the initial values of the rotor speed and the generator rotor speed, respectively. The energy conservation before and after the grid fault is applied:

$$(W+E)_{BEFORE} = (W+E)_{AFTER} \ , \tag{5.13}$$

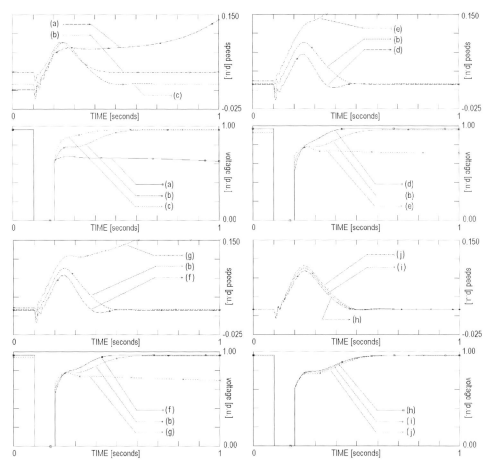

Figure 5.6 Dynamic behaviour of speed and terminal voltage influenced by generator parameters: (a) $-R_R \frac{1}{2}$, $X_R 1$, $X_S 1$, $R_S 1$, instability, (b) $-R_R 1$, $X_R 1$, $X_S 1$, $R_S 1$, recovery, (c) $-R_R 2$, $X_R 1$, $X_S 1$, $R_S 1$, recovery, (d) $-R_R 1$, $X_R \frac{1}{2}$, $X_S 1$, $R_S 1$, recovery, (e) $-R_R 1$, $X_R 2$, $X_S 1$, $R_S 1$, collapse, (f) $-R_R 1$, $X_R 1$, $X_S \frac{1}{2}$, $R_S 1$, recovery, (g) $-R_R 1$, $X_R 1$, $X_S 2$, $R_S 1$, collapse, (h) $-R_R 1$, $X_R 1$, $X_S 1$, $R_S \frac{1}{2}$, recovery, (i) $-R_R 1$, $X_R 1$, $X_S 1$, $R_S 1$, recovery, (j) $-R_R 1$, $X_R 1$, $X_S 1$, $R_S 2$, recovery.

Additional assumptions are made according to (Akhmatov et al., 2000(a)).

1) During the fault, the shaft relaxes completely, implying that the shaft twist is reduced to zero.
2) The rotor inertia constant is much larger than the generator rotor inertia constant. When the grid fault duration is short enough, the rotor speed does not change significantly during the fault.
3) Only the generator rotor speed will change during the fault, and the generator rotor speed is increased by $\Delta \omega_G$.

Using the above assumptions, energy conservation Eq.(5.13) leads to:

$$H_M \cdot \omega_M^2 + H_G \cdot \omega_G^2 + \tfrac{1}{2} \cdot K_S \cdot \theta_S^2 = H_M \cdot \omega_M^2 + H_G \cdot \left(\omega_G + \Delta\omega_G\right)^2 + 0. \qquad (5.14)$$

When combining Eq.(5.11) with Eq.(5.14), the increase of the generator rotor speed due to the shaft relaxation will be:

$$\Delta\omega_G \sim \frac{T_M^2}{H_G K_S}. \qquad (5.15)$$

This relationship demonstrates that shaft relaxation contributes to the acceleration of the generator rotor when the power grid is subject to a short-circuit fault. Eq.(5.15) shows also that the lower the shaft stiffness is, the larger the expected acceleration of the generator rotor due to shaft relaxation during the grid fault.

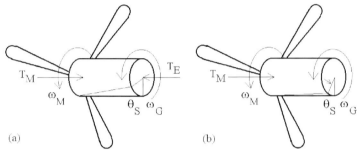

Figure 5.7　Operation of the shaft system of the wind turbine during a grid fault: **(a)** - pre-twisted in normal operation of the grid, **(b)** - relaxation (reduction of the shaft twist, θ_S) when the fault occurs.

Figure 5.8 compares the generator rotor speed of an induction generator with the generator rotor inertia constant H_G =0.5 s to the generator rotor speed of the wind turbine with the inertia constants H_M =2.5 s and H_G =0.5 s, and the shaft stiffness K_S =0.3 p.u./el.rad. The same figure shows also the rotor speed of the wind turbine. As can be seen, the assumption that the rotor speed, ω_M, does not change much during a short duration fault is correct. The generator rotor is accelerated more in the case of the wind turbine (with the shaft relaxation) than in the case of the induction generator alone (without the shaft relaxation). The shaft as a two-mass model is relevant in investigations of power system stability because this may lead to slowing the voltage recovery rate.

Figure 5.9 compares the simulated voltage curves of windfarm 1 commissioned in the power grid shown in **Figure 5.3**. The short-circuit fault has a duration of 100 ms. When the windfarm is modelled using a two-mass model regarding the shaft relaxation, the voltage recovers much slower than in the case of a lumped-mass model assuming an ideally stiff shaft.

Understanding the shaft relaxation mechanism is important to improve the ride-through capability of fixed-speed wind turbines (Akhmatov et al., 2000(a); Salman and Teo, 2003).

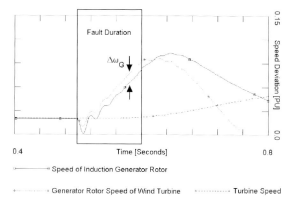

Speed of Induction Generator Rotor

Generator Rotor Speed of Wind Turbine   Turbine Speed

Figure 5.8   More acceleration of the generator rotor due to shaft relaxation during a grid fault. Reprinted from Ref. (Akhmatov, 2003(b)), Copyright (2003), with permission from the copyright holder.

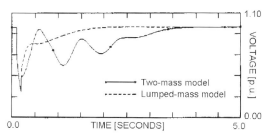

Figure 5.9   Shaft relaxation process slow the voltage recovery rate.

5.6   Application of dynamic stability limit

Instability of grid-connected induction generators, as well as wind turbines equipped with such induction generators is caused by excessive over-speeding of the wind turbines during grid faults. Definition of the dynamic stability limit is therefore relevant to technical arrangements and control of the power grid and of the wind turbines to improve short-term voltage stability and fault-ride-through capability. With respect to the dynamic stability limit, avoiding excessive over-speeding, e.g. keeping the rotor speed below the critical speed, ω_{CR}, may improve the stability and the ride-through capability of the fixed-speed wind turbines.

5.6.1   What kind of a fault gives the worst case?

Excessive overspeeding of induction generators is critical for maintaining voltage stability and may lead to disconnection of the generators from the grid. The worst case with regard to the fault-ride-through capability of such induction generators can, then, be evaluated from the generator rotor speed during the fault. In the following, the speed behaviour is evaluated with regard to different kinds of short-circuit faults:

1) A single-phase short-circuit fault (1p) to the ground.

2) A 2-phase short-circuit fault (2p) phase-to-phase.

3) A 2p fault to the ground.

4) A 3-phase short-circuit fault (3p).

Simulation results for generator rotor speed and terminal voltage of a 500 kW induction generator in rated operation are shown in **Figure 5.10**. As can be seen from the speed plots, the worst case
with regard to overspeeding occurs in the case of the 3-phase short-circuit fault, which also presents
the worst case with regard to the voltage stability and the fault-ride-through capability of the induction generators.

Other kinds of faults, i.e. the single-phase to ground and the phase-to-phase faults, do not result
in such intense overspeeding of the induction generators. As known from practical operation in
Denmark, fixed-speed wind turbines equipped with such induction generators ride through single-
phase and phase-to-phase faults in the grid, which are also the most frequent types of faults. If induction generators may ride through a 3-phase short-circuit fault at a given node of the transmission
power system, then the induction generators will also ride through all other kinds of short-circuit
faults (of the same duration) subjected to the same node.

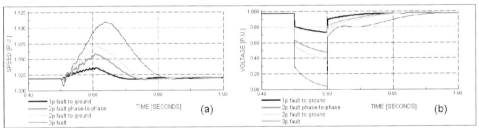

Figure 5.10   Behaviour of: **(a)** - generator rotor speed and **(b)** - terminal voltage of the 500 kW induction generator
during different kinds of short-circuit faults. The 3-phase short-circuit fault results in the largest overspeeding of the generator compared to all the other kinds of faults.

From this point of view, investigations of short-term voltage stability and fault-ride-through capability of fixed-speed wind turbines can be limited to the investigations with such 3-phase short-
circuit faults, e.g. investigations with other kinds of faults will not be required. Therefore the following presentation for fixed-speed wind turbines is made with regard to 3-phase short-circuit
faults.

5.7   Ride-through capability of large windfarms

Wind technology applied to large (offshore) windfarms must be robust, tried and tested. Fixed-
speed wind turbines with induction generators have been in operation on sites in Denmark for many
years. Denmark has now two large offshore windfarms with fixed-speed, active-stall controlled
wind turbines equipped with induction generators (Sørensen et al., 2001). The Middelgrund offshore windfarm consists of twenty 2 MW wind turbines from the manufacturer Siemens. The Middelgrund offshore windfarm was commissioned in 2001 and connected to the 30 kV Copenhagen

distribution power grid. The Middelgrund offshore windfarm is shown in **Figure 5.11** and has become the badge of Copenhagen.

The Rødsand 1 / Nysted offshore windfarm has a rated power capacity of 165 MW divided between 72 wind turbines from the manufacturer Siemens. This offshore windfarm was commissioned in 2003 and sited south of the island of Lolland, just south of the main Danish island of Zealand. The location of this offshore windfarm containing fixed-speed wind turbines is marked on the map shown in **Figure 5.12**. The Rødsand 1 / Nysted offshore windfarm feeds into the relatively weak Lolland power system and is remote from the main consumption centre of Copenhagen (in north outside the map). The short-circuit capacity in this region is in the range of 1000 MVA to 1800 MVA depending on the connection of the transmission system in the vicinity of the connection point of the Rødsand 1 / Nysted offshore windfarm.

The Danish transmission system operator has formulated the Grid Code[9] addressed to the windfarms connected to the transmission power system (with a voltage above 100 kV). The Danish Grid Code requires that the voltage and frequency in the transmission power system are re-established without subsequent disconnection of the large windfarms (Energinet.dk, 2004(b)). In other words, the Danish Grid Code requires a fault-ride-through capability of the wind turbines commissioned in large windfarms. Similar requirements dealing with a fault-ride-through capability are present in the national Grid Codes of many countries, especially as the amount of wind power increases rapidly around the globe.

Figure 5.11 The Middelgrund offshore windfarm, Copenhagen, Denmark. Photo copyright Siemens. Reproduced with permission from Siemens WPG.

---

[9] Note that the Rødsand 1 / Nysted wind farm is subject to the Grid Code (Eltra, 2000) as it was commissioned before 2004.

The ride-through capability of large windfarms must be demonstrated by test and by simulations. Below simulations are made using detailed models of large windfarms containing many wind turbines.

5.7.1 Modelling a large windfarm

The fault-ride-through capability of a large offshore windfarm will now be investigated via the use of a detailed model of an offshore windfarm consisting of eighty 2 MW wind turbines. The investigations are made using the simulation tool PSS/E^{TM} (Akhmatov et al., 2003(a)). Every wind turbine within this windfarm is represented using a dynamic model of a fixed-speed, active-stall controlled wind turbine equipped with a no-load compensated induction generator. As the base case, the rotor winding resistance is R_R =0.020 p.u., the generator rotor inertia constant is H_G =0.5 s, the rotor inertia constant is H_M =2.5 s, and the shaft stiffness is K_S =0.3 p.u./el.rad.

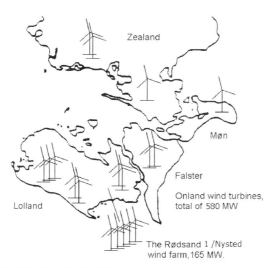

Figure 5.12 Location of the Rødsand 1 / Nysted offshore windfarm in Eastern Denmark. Around 580 MW wind power is incorporated within the same geographical area on-land (the year 2005).

Through 0.7/30 kV transformers, the induction generators of the wind turbines are connected to the internal windfarm network. The internal network is arranged in eight rows (eight sections) with ten electricity-producing wind turbines in each section. Within the rows, the wind turbines are connected to a 30 kV sea-cable. The distance between two neighbouring wind turbines within the same section is 500 m and the distance between two neighbouring sections is 850 m. Such distances are chosen to comply with an empirical role of the five rotor diameters. This ensures that any shadow effect is significantly reduced. The 30 kV sea-cable sections are connected to an offshore platform with a 30/30/132 kV tertiary transformer. Through a 132 kV sea- and underground- cable, the offshore windfarm is then connected to the connection point in the on-land transmission system. In this case, an AC- connection to the transmission network is chosen, see **Figure 5.13.**

The large offshore windfarm is set to be reactive neutral with the power grid at the connection point. This implies that the reactive power exchange between the windfarm and the power grid must be around zero independently from the operational point of the offshore windfarm. This requires that a number of capacitors and reactors are incorporated at the connection point or at the offshore platform.

The detailed model allows simulation of the offshore windfarm on the assumption of an irregular wind distribution. This assumption is realistic because the wind turbines within the offshore wind-farm are shadowing each other from incoming wind. Furthermore, the area of the windfarm is 4.5 x 6 km^2 which gives rise to expected irregular wind distribution over such a large area.

The power generation pattern shown in **Figure 5.13** is according to the wind distribution over the offshore windfarm assumed in this investigation. At the given wind distribution, the efficiency of the windfarm is 93%, supplying approximately 150 MW to the power grid. The short-circuit capacity of the transmission system at the connection point is 1800 MVA. A short circuit fault is subject to a selected node in the transmission system, with a duration of 150 ms. When the short-circuit fault is cleared, the transmission lines connected to the faulted node are permanently disconnected. Disconnection of the transmission lines leads to a reduction of the short-circuit capacity of the transmission system, to 1000 MVA.

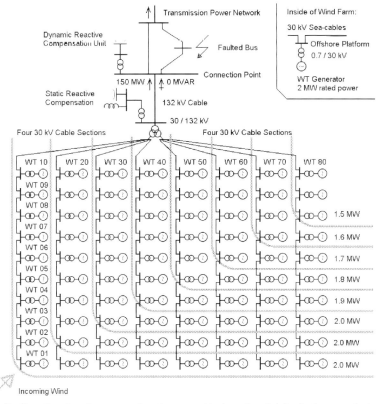

Figure 5.13   Configuration of, and power supply pattern caused by irregular wind distribution over, the large offshore windfarm. Reprinted from Ref. (Akhmatov et al., 2003(a)), Copyright (2003), with permission from Elsevier.

Before the investigations with the short-circuit fault are started, it is ensured that the line disconnection itself (reduction of the short-circuit capacity to 1000 MVA) does not lead to voltage instability in the grid. If voltage instability occurs, this is to be initiated by the short-circuit fault subject to the selected node of the transmission system.

5.7.2   Why the detailed model is chosen

In this investigation, this complete representation of the offshore windfarm is chosen to answer the commonly asked question if there can be power oscillations between the fixed-speed wind turbines closely located to each other in a large windfarm (Akhmatov et al., 2003(a))? Such power oscillations could be started by a grid disturbance. The concern about possible power oscillations is enforced when wind turbines are at different operational points, equipped with relatively soft shafts and equipped with control systems, such as with blade-angle control. Furthermore, the induction generators of the wind turbines experience a little different short-circuit capacities from the power grid. The power oscillations between the wind turbines, if these have occurred, would threat the fault-ride-through capability of the windfarm.

The fault-ride-through capability of the windfarm is based on the two issues that must be fulfilled.

1) The grid voltage must re-establish when the short-circuit fault is cleared.
2) The protective relay settings are adjusted to avoid an unnecessary tripping of the wind turbines during the short-circuit fault (Akhmatov et al., 2001).

All the simulations are first made on the assumption that the protective relays do not trip the wind turbines so long as the grid voltage recovers. The relay settings, and the part of the disconnected wind power will be evaluated later in **Section 5.8**.

5.7.3   Case of voltage instability

First and foremost, it is necessary to investigate if a short-circuit fault in the transmission power system may cause voltage instability of the large (offshore) windfarm. Presumably, such undesired situations may occur when no arrangements which would improve voltage stability or ride-through capability of the wind turbines are applied.

Modern fixed-speed wind turbines are equipped with active-stall control. Regular active-stall control is developed and applied to optimise the power output of the rotor according to incoming wind. Usually, regular active-stall control disregards the concerns of short-term voltage stability. An example of the regular active-stall control is when the control signal is in the form of the electric power, $X = P_E$, and the reference signal is the reference power, $X_{REF} = P_{REF}$. The reference power is defined in accordance to the incoming wind, V, (Hinrichsen, 1984; Akhmatov, 2001).

When the transmission power system is subject to a short-circuit fault, the voltage, U_S, and then the active power, P_E, supplied from the induction generator drop. The regular active-stall control will, presumably, interpret this as a lack of power output and try either to increase, or to maintain, the mechanical power (either keeping β_{REF} at β_{OPT} if it is already at the optimised position or tracking β_{REF} to β_{OPT} if the reference angle is not optimal).

When keeping the reference pitch angle at the optimised position, the pitch angle, β, is kept fixed as before the grid fault occurred. This action corresponds to fixed-pitch operation as in the case of stall-controlled wind turbines. This control action may be present when the wind speed is above the rated value. Tracking the reference pitch angle, β_{REF}, and the actual pitch angle, β, to the optimised position, β_{OPT}, may lead to ever larger acceleration of the rotor than before the grid fault occurrence (Akhmatov, 2001). This control action may be present when the wind speed is below the rated wind speed.

In investigations of short-term voltage stability, fixed-speed active-stall wind turbines can therefore be represented as stall-controlled. When neither dynamic reactive compensation nor other controls are applied, a short-circuit fault will result in a voltage instability of the large offshore windfarm. The simulation curves are shown in **Figure 5.14**. So long as the wind turbines maintain grid connection, the fluctuations of the generator rotor speed, the terminal voltage and the other parameters registered at the different wind turbines within the large offshore windfarm are "in-phase" and not against each other. Such "in-phase" fluctuations are denoted as a coherent response. The natural frequency of the fluctuations is equal to the shaft torsion mode.

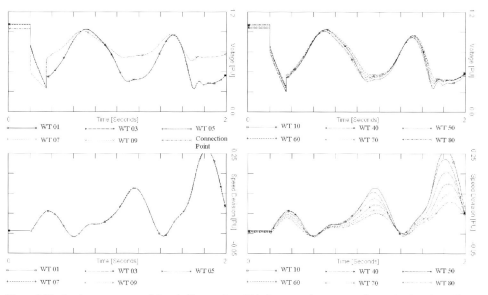

Figure 5.14 A coherent response of the windfarm to the grid fault – excessive over-speeding and voltage instability of the windfarm when no control arrangements are applied. Reprinted from Ref. (Akhmatov et al., 2003(a)), Copyright (2003), with permission from Elsevier.

Note that such a voltage instability will probably not develop into a voltage collapse. The wind turbines are equipped with protective relays. When an abnormal grid operation such as an uncontrollable voltage decay is registered, the wind turbines will disconnect from the power grid. When the excessively over-sped wind turbines have disconnected, the grid voltage re-establishes. The subsequent disconnection of the large offshore windfarm requires establishment of immediate power reserves. In this particular case, immediate power reserves of 150 MW will be required if the whole windfarm trips out.

Voltage instability, causing protective disconnection of the windfarm, implies however that the wind technology and the technical solution arranged for the windfarm do not comply with the national Grid Code (Eltra, 2000).

5.7.4 Dynamic reactive compensation

The use of dynamic reactive compensation may improve voltage stability of the power grid and fault-ride-through capability of a large offshore windfarm (Eriksson et al., 2003). Dynamic reactive compensation can be arranged in several ways. The choice of the arrangement depends on the demands for reactive power to get the grid voltage re-established, the controllability of the equipment and the cost. In terms of the dynamic stability limit, the use of dynamic reactive compensation implies "lifting-up" the electrical torque versus speed characteristic in the speed range above the rated speed. This can be understood when considering that the electrical torque versus speed characteristic of the induction generator, shown in **Figure 5.1**, will follow the curve computed for U_S =1.0 p.u. in the speed range above the kip-speed. This arrangement leads to a greater critical speed, ω_{CR}, and hence to improvement of voltage stability.

The target, however, is to re-establish the grid voltage and to get the large windfarm to ride through the short-circuit fault in the transmission power system. The grid voltage must be re-established within a defined operational range. This implies that the voltage may not show a tendency to uncontrollable decay. Excessive over-voltage would not be acceptable either, as this may cause other damage. Therefore dynamic reactive compensation required for the windfarm may be capacitive as well as inductive.

So long as protective disconnection of the fixed-speed wind turbines is not used, the induction generators of such wind turbines absorb reactive power from the grid during voltage re-establishment. Therefore the character of the dynamic reactive compensation required in this case will be capacitive, and dynamic reactive compensation units outlined by (Cigré, 1999; Kundur, 1994; Noroozian et al., 2000; Barocio and Messina, 2003) can be used. These include, but are not restricted to:

1) A Synchronous Compensator, which is a synchronous machine running without a prime mover or a mechanical load. The reactive power output of this unit is controlled by the field excitation. With the use of a voltage regulator, Synchronous Compensators can automatically adjust reactive power exchanged with the grid to maintain a desired terminal voltage.

2) A Static VAR compensator (SVC) contains a number of shunt-connected capacitors generating, and reactors absorbing, reactive power and whose outputs are co-ordinated. The term "static" indicates that the device contains no moving or rotating components. The switching of the capacitors and continuous control of the reactors are arranged by thyristor switches, respectively, thyristor valves.

3) A Statcom or a Static Synchronous Compensator is a voltage source converter controlled by power electronics with Insulated Gate Bipolar Transistor (IGBT) -switches (Søbrink et all, 1998, Wu et al., 2003), (Cigré, 1999).

Depending on the actual demands for dynamic reactive power control, the compensation unit can be arranged with the use of continuous, discrete or a combination of both kinds of control (Kundur, 1994; Akhmatov and Nielsen, 2005). Synchronous Compensators and Statcoms use continuous control. Synchronous Compensators are characterised by relatively high running and maintenance costs because of the presence of rotating and moving parts (Taylor, 1994), whereas Statcoms are relatively expensive features, but have lower running costs. The advantage is, however, that Synchronous Compensators and Statcoms are able to set voltage at their terminals and restore voltage in the power grid after a collapse.

SVC units need grid voltage to control the reactive power in the grid. Being arranged with the use of passive components, the SVC units cannot restore voltage in the power grid after a collapse. The control of the SVC unit can be set up in several ways depending on the technical requirements and the desired cost.

1) Continuous control arranged with thyristor-controlled reactors in series with fixed capacitors. A SVC unit with continuous control only is an efficient, but relatively expensive feature.
2) Discrete control using a number of thyristor- or mechanically switched capacitors and reactors. This is a relatively cheap.
3) A combination of discrete and continuous control within the same SVC unit. Here the thyristor controlled reactor is combined with a number of thyristor- or mechanically switched capacitors. In this solution, the demands for the reactive power control and the cost are optimised knowing the actual demands for control from investigations of power system stability.

It can be advantageous to apply control of the SVC unit to reach fixed capacitors and reactors which already are installed in the power grid near the unit's connection point. When applying the control system of the SVC unit to switch the exciting capacitors and reactors, the range of the reactive power control can be expanded and more efficient. This may also reduce the size and cost of the SVC unit itself. The running costs of SVC units are lower than the running costs of Synchronous Compensators, but continuously controlled SVC units can be more expensive to establish than Synchronous Compensators.

In Ref. (Akhmatov, 2003(b)), the demands for dynamic reactive compensation to re-establish the voltage in the large offshore windfarm are examined. Following results are presented for the cases of a SVC unit with continuous control, Synchronous Compensators with different control systems and, in short, a Statcom.

5.7.4.1 Static VAR Compensator

The large offshore windfarm is equipped with fixed-speed, active-stall controlled wind turbines. The active-stall control is only applied to optimise power output of the wind turbines and do not support fault-ride-through capability of the wind turbines.

To improve the voltage stability and the fault-ride-through capability of the large offshore windfarm, the SVC unit with continuous control is applied. The dynamic model of the SVC and its continuous control is described and validated in Ref. (Noroozian et al., 2000). In other words, the SVC unit and its control are represented with the use of the component-oriented model of the ABB equipment. The investigation parameter is the reactive power capacity of the SVC unit.

It is found that the voltage re-establishes when the reactive power capacity of the SVC unit is at least 100 MVAr. When the reactive power capacity of the SVC unit is less than 100 MVAr, the voltage instability occurs and the large offshore windfarm disconnects from the grid. The simulated curves of the voltage and the generator rotor speed of the selected wind turbines within the large offshore windfarm are given in **Figure 5.15** and **Figure 5.16**.

The electrical and mechanical parameters of the wind turbines such as the voltage, speed, active and reactive power show fluctuating behaviour initiated by the short-circuit fault. The wind turbines show a coherent response to the grid fault, and there is no indication that the dynamic control of the SVC will interact with the active-stall control of the wind turbines.

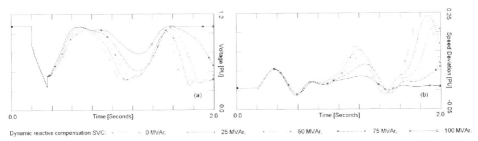

Figure 5.15   Reactive power capacity of the SVC unit versus voltage stability: **(a)** – terminal voltage of the wind turbine WT 01, **(b)** – generator rotor speed of WT 01. Reprinted from Ref. (Akhmatov, 2003(b)), Copyright (2003), with permission from the copyright holder.

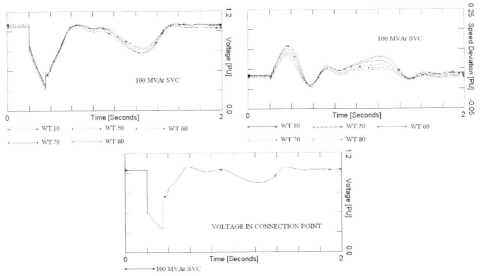

Figure 5.16   Voltage and generator rotor speed show a coherent response when the SVC unit of 100 MVAr is applied. The voltage is re-established when the short-circuit fault is cleared. Reprinted from (Akhmatov et al., 2003(a)), Copyright (2003), with permission from Elsevier.

5.7.4.2 Synchronous Compensator

Synchronous Compensators can be established as new equipment or arranged as synchronous generators and their excitation control of the power plants taken out of service (Akhmatov, 2003(b)). Synchronous generators and their excitation control can, then, be relocated to the relevant nodes of the power grid, for example to a connection point of a large windfarm.

Excitation control of the Synchronous Compensators applied in this investigation is by the control of two Danish power plant units at Asnæsværket denoted as ASV1 and ASV3. Two different control systems are investigated to illustrate that the windfarm response depends on the reactive power control. Simulation models of the generators and their excitation control of ASV1 and ASV3 are kindly offered by the Danish power company NESA. Thus, the investigation results are obtained using validated models of the reactive power control applicable to the Danish offshore windfarms.

Figure 5.17 shows the voltage profiles in the large windfarm when the Synchronous Compensator is applied to stabilise operations. As can be seen, the voltage profiles depend on the control of the Synchronous Compensator. The voltage recovers faster when the control of ASV1 is applied. This is explained by the fact that the control of ASV1 reacts faster on changes of grid voltage than that of ASV3. This demonstrates that voltage recovery depends not only on reactive power capacity, but also on the control of the dynamic reactive compensation unit.

The reactive power capacity of the Synchronous Compensator is to be in the same range as that of the SVC unit to get the grid voltage to re-establish correctly (Akhmatov et al., 2001). In this case, this is at 100 MVAr.

The voltage drop at the wind turbine terminals, as well as the acceleration of the wind turbine generators during the grid fault, is lower when using the Synchronous Compensator than when using the SVC unit. This difference is caused by the fact that the Synchronous Compensator starts already to supply the reactive power when the short-circuit fault is subject to the power grid, as shown in **Figure 5.17(f)**, e.g. providing a short-circuit capacity to the grid. The Synchronous Compensator generates a desired voltage at the terminals which characterises operation of synchronous generators (with their excitation control).

In contrast to the Synchronous Compensator, the SVC unit starts to supply reactive power when the fault is removed, as shown in **Figure 5.17(f)**. The SVC unit operates as the reactive power source when the grid voltage stays above a given limit. When the grid voltage drops below this limit, the SVC unit may reach its boost limit. At the boost limit, operation of the SVC unit can be compared to operation of a shunt-capacitor with reactive power $Q_C = X_C U_S^2$ (Taylor, 1994). In this way, reactive power generation decreases as the grid voltage squared.

When the Synchronous Compensator is applied to stabilise operation and improve ride-through capability of the large offshore windfarm, the electrical and mechanical parameters of the wind turbines show a coherent response. The voltage and the generator rotor speed of the different wind turbines oscillate "in-phase" and do not interact with the dynamic excitation control of the Synchronous Compensator.

5.7.4.3 Application of Statcoms

When the Statcom is applied for dynamic reactive compensation of the large offshore windfarm, the reactive power capacity of the Statcom must be at least 75 MVAr. This is less than the minimal demands for the reactive power capacity of the SVC and of the Synchronous Compensator.

However the Statcom can be the most expensive (per 1 MVAr) system compared to SVC and Synchronous Compensators. Reduction of the reactive power capacity is not by itself a sufficient argumentation for choosing the Statcom instead of the SVC or the Synchronous Compensator in the case of the given offshore windfarm. The running cost of the SVC and the Statcom are considered similar since this equipment is without rotating or moving parts, but controlled by power electronics (Taylor, 1994; Cigré, 1999).

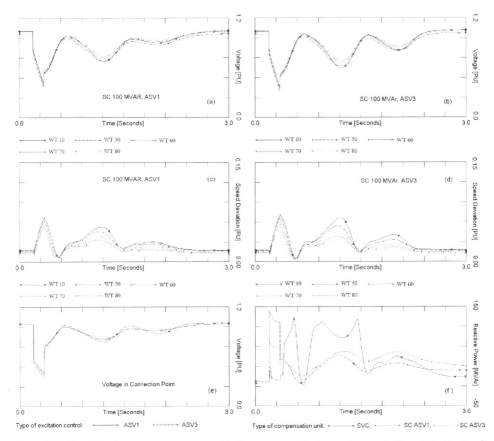

Figure 5.17   Use of Synchronous Compensator to stabilise operation of the large offshore windfarm: **(a)**, **(b)** – terminal voltage of selected wind turbines applying the excitation control of ASV1, respectively, of ASV3, **(c)**, **(d)** – generator rotor speed of selected wind turbines applying the excitation control of ASV1, respectively, of ASV3, **(e)** – voltage in connection point applying the excitation control of ASV1, respectively, of ASV3, **(f)** – reactive power supplied to the grid from the compensation unit. Reprinted from Ref. (Akhmatov, 2003(b)), Copyright (2003), with permission from the copyright holder.

For rounding up the discussion about choosing a dynamic compensation unit for a large offshore windfarm, it has been decided to incorporate a SVC unit at the 132 kV substation Radsted, which is the connection point of the Rødsand 1 / Nysted offshore windfarm (165 MW rated power, with fixed-speed wind turbines) to reduce undesired voltage fluctuations in the near power grid (Rasmussen et al., 2003). The SVC unit is the first of its kind in the Danish power system and financed from the Danish Public Service Obligation (PSO) fund as part of a research program which shall demonstrate operation of dynamic reactive compensation together with this large offshore windfarm. The contractor is the power distribution company SEAS-NVE. The SVC unit will be delivered from the manufacturer Siemens PTD and have a rating of 85 MVAr.

5.7.5   Influence from wind turbine construction and control

The fault-ride-through capability of wind turbines can be improved by the construction and the control of fixed-speed wind turbines. Such improvements are based on the dynamic stability limit of the fixed-speed wind turbines and may lead to reductions in the capacity demands for dynamic reactive compensation. The reduction of the capacity demands for dynamic reactive compensation is therefore applied as the measure of the construction improvements and the control which are dedicated to improve the fault-ride-through capability of the fixed-speed wind turbines.

5.7.5.1   Generator parameters

In **Section 5.3.2**, it is explained that voltage stability of induction generators depends on parameters such as the stator resistance, stator reactance, magnetising reactance, rotor resistance and rotor reactance. This dependence is explained in Ref. (Akhmatov et al., 2000(c)) in terms of increasing the critical speed when reducing the stator resistance, stator reactance, magnetising reactance and rotor reactance and when increasing the rotor resistance. Such induction generator parameters are given by the generator design (not to alter). However, the total resistance of the rotor circuit can be increased due to a chosen design. In this section, improvement of voltage stability is demonstrated in simulations for increased rotor resistance, R_R.

Increasing the critical speed of fixed-speed wind turbines by increasing the rotor resistance, R_R, by the ratio of 2 is illustrated in **Figure 5.18**. In this case, the total rotor resistance $R_R = 2 R_{R0} = 0.040$ p.u. and the critical speed, ω_{CR}, is significantly increased compared to the case of the default generator parameters. Increasing the critical speed leads to improvements of voltage stability and reduction of capacity demands for dynamic reactive compensation (Akhmatov et al., 2003(a)).

When the rotor resistance is increased, i.e. $R_R = 2 R_{R0} = 0.040$ p.u., the capacity demands for the continuously controlled SVC unit are reduced to 25 MVAr. Simulation curves for the windfarm using a 25 MVAr SVC unit are shown in **Figure 5.19**. These demands must be compared to 100 MVAr required in the case of the default generator data with rotor resistance of $R_{R0} = 0.02$ p.u.

The capacity demands for dynamic reactive compensation are significantly reduced when the rotor resistance is doubled. On the other hand, this feature leads to increasing the power losses in the rotor circuit accordingly to $P_R = R_R I_R^2$. The use of generators with an increased rotor resistance may require a cooling system applied to the generator rotor when the rated power of the wind turbine is in the MW range.

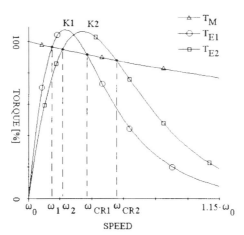

Figure 5.18 Dynamic stability improvement by increasing the rotor resistance, R_R, where T_M – the mechanical torque, T_{E1} – the electrical torque with $R_R=R_{R0}$, and T_{E2} – the electrical torque with $R_R=2R_{R0}$. When $R_R=R_{R0}$, the wind turbine can operate within the speed range from its initial speed, ω_1, to the critical speed, ω_{CR1}, without loss of dynamic stability. When the rotor resistance is $2R_{R0}$, the stable operation range is expanded to be from the initial speed, ω_2, to the critical speed, ω_{CR2}. $K1$ and $K2$ are marking the kip-torque of the given induction generators, respectively. Reprinted from (Akhmatov et al., 2003(a)), Copyright (2003), with permission from Elsevier.

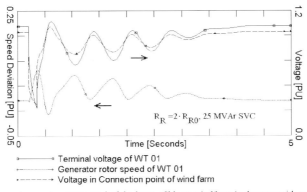

Figure 5.19 Voltage and generator rotor speed of the large offshore windfarm in the case with an increased rotor resistance and reduced capacity of dynamic reactive compensation. Reprinted from (Akhmatov et al., 2003(a)), Copyright (2003), with permission from Elsevier.

5.7.5.2 Mechanical construction

The mechanical construction parameters are first and foremost the rotor inertia constant, H_M, and the shaft stiffness, K_S. Such mechanical parameters of fixed-speed wind turbines may influence the capacity demands for dynamic reactive compensation, voltage stability of the power grid and fault-ride-through capability of the wind turbines themselves.

A common opinion has been that the higher inertia of the rotor is, the more stable the operation of wind turbines expected when the power system is subject to short-circuit faults. In terms of the dynamic stability limit, defined in **Section 5.4**, the rotor inertia does not influence the critical speed of wind turbines. This implies that two wind turbines with identical generator parameters, but with different inertia constants H_{M1} and H_{M2}, where $H_{M1} > H_{M2}$, have the same critical speed $\omega_{CR1} = \omega_{CR2}$. Since the rotor inertia constants are different, the wind turbines will accelerate differently when the power grid is subject to a short-circuit fault. Thus, these two wind turbines will have different critical failure times, t_{CR}, defined as the time of the grid fault where the wind turbine still maintains in stable operation after the fault is cleared. Obviously, the heaviest wind turbine has the largest critical failure time and therefore $t_{CR1} > t_{CR2}$. Therefore heavy wind turbines show more stable behaviour compared to light wind turbines, so long as the failure time is not excessive. In practical situations, the failure time is relatively short (from 100 ms to 250 ms) and heavy wind turbines will be preferred with regard to voltage stability and better fault-ride-through capability (Akhmatov et al., 2000(c)).

Since shaft systems are characterised by relatively low shaft stiffness, K_S, the shaft systems accumulate an amount of potential energy when the wind turbines and the power grid are in normal operation (Akhmatov et al., 2000(b)). In **Section 5.5** it is demonstrated that the lower shaft stiffness is, the more potential energy is accumulated in the twisted shafts. When the power grid is subject to a short circuit fault, the twisted shafts start to relax. The potential energy is transformed into generator rotor kinetic energy. The shaft relaxation during the grid fault results in a more intense acceleration of the generator rotor.

Section 5.5 has shown that the contribution from the shaft system to a more intense acceleration of the generator rotor is opposite-proportional to, the shaft stiffness, K_S. Therefore the increase in shaft stiffness, K_S, leads to a reduction of the wind turbine over-speeding and improvement in voltage stability and fault-ride-through capability. This is illustrated by simulations in **Figure 5.20**.

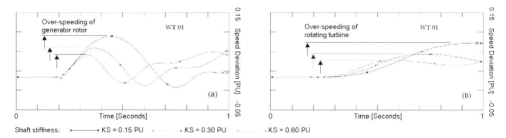

Figure 5.20 Wind turbine over-speeding caused by a short-circuit fault, at fixed H_M=2.5 s and H_G=0.50 s and varying shaft stiffness K_S: **(a)** – generator rotor speed, **(b)** – rotor speed. No dynamic reactive compensation is used in these simulations. Reprinted from (Akhmatov et al., 2003(a)), Copyright (2003), with permission from Elsevier.

The increase in shaft stiffness and rotor inertia corresponds to enforced mechanical constructions of the wind turbines. For instance, the shaft stiffness can be increased by making the shafts shorter and thicker. The rotor inertia can be increased by making the blades heavier and longer.

Simulation results showing the reduction of the capacity demands for dynamic reactive compensation depending on the mechanical construction parameters H_M and K_S are collected in **Table 5.1**.

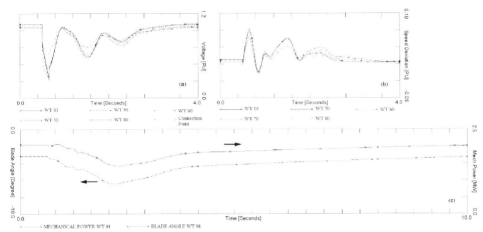

Figure 5.22   Voltage re-establishment in the large offshore windfarm with fixed-speed active-stall control wind turbines applying the generator rotor speed as the input signal: **(a)** – terminal voltage of selected wind turbines of the windfarm and in the connection point, **(b)** – generator rotor speed of selected wind turbines,
(c) – mechanical torque and the blade angle of the wind turbine WT 01. No dynamic reactive compensation
is used to re-establish the grid voltage. Reprinted from (Akhmatov et al., 2003(a)), Copyright (2003), with
permission from Elsevier.

Again, the grid voltage is not a parameter which can be applied to optimise the power output of
wind turbines. Therefore, the input signal to the regular active-stall control system contains the active power of the wind turbine. The reference signal applied is $X_{REF} = P_{REF}$ whereas the input control
signal is $X = P_E/U_S^2$. This input signal is suggested in (Akhmatov et al., 2003(a)) because of the following:

1) Optimisation of the output power of the wind turbine will be gained when the power grid is at
   normal operation (remember the dead-band in the measured voltage signal).
2) When the voltage is below U_{DMIN}, this indicates that the power grid can be subject to a grid
   fault and the wind turbines must reduce the mechanical torque to improve the fault-ride-
   through capability.
3) The voltage in the denominator, U_S, is squared because the active power, P_E, is proportional
   to the terminal voltage squared, Eq.(5.1) and Eq.(5.2). In this way, the error signal to the regular active-stall control, X_{ERR}, is compensated for the voltage drop during the short-circuit fault.

The simulation results for the large offshore windfarm when the regular active-stall control applies an additional voltage signal to produce the input signal, $X = P_E/U_S^2$, are shown in **Figure 5.23**.
When this active-stall control is applied, the grid voltage is re-established without use of any dynamic reactive compensation. An active-stall control has, therefore, significantly improved the voltage stability and the ride-through capability of the large offshore windfarm.

Accurate tuning of the control system is essential in such investigations of power system stability. With the parameters of the regular active-stall control applied in this investigation, neither
power oscillations between the different wind turbines in the large windfarm, nor interaction be-

tween the control systems of the wind turbines, have been seen. The wind turbines show a coherent response to the grid fault.

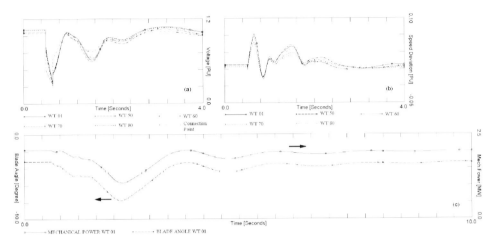

Figure 5.23   Voltage re-establishment in a large offshore windfarm using the regular active-stall control and an input signal containing the terminal voltage: **(a)** – terminal voltage of selected wind turbines and in the connection point, **(b)**   generator rotor speed of selected wind turbines, **(c)** – mechanical power and the blade angle of wind turbine WT 01. No dynamic reactive compensation is applied to re-establish the grid voltage. Reprinted from (Akhmatov et al., 2003(a)), Copyright (2003), with permission from Elsevier.

5.7.5.5   Blade angle control – Power ramp

To stabilise the large offshore windfarm and contribute to the fault-ride-through capability, the wind turbines can be ordered to reduce their mechanical power from an arbitrary operational point down to 20% of the rated power in less than 2 seconds. This ordered power reduction over a given time, is called the power ramp. The technical requirement for the use of the power ramp at the grid disturbances was formulated by the Danish system operator in the year 2000 with regard to commissioning the first offshore windfarms in Denmark (the Horns Rev A and the Rødsand 1 / Nysted windfarms) (Eltra, 2000).

The order to apply the power ramp is given by an external signal sent from the external system to the large offshore windfarm. For instance, the external system monitors the grid voltage at the connection point of the windfarm. The external signal is sent when the voltage at the connection point of the windfarm suddenly drops below a given value (Akhmatov et al., 2003(a)). A delay is introduced from the moment when the grid fault occurs to the moment when the external system registers the fault event and sends the external signal to apply the power ramp. In this investigation, this delay is set to 200 ms. Note that the delay is longer than the fault duration. This implies that the external signal to apply the power ramp is first sent after the grid fault is removed.

When the power ramp is completed, the large offshore windfarm continues to operate at 20% of its rated power so long as the order is given. When the voltage and the frequency are re-established in the power grid, the external system cancels the power ramp order. From this moment, the large offshore windfarm re-establishes its normal operation.

generators (Akhmatov et al., 2003(a)). In the absence of synchronising torque, fluctuations of the electrical and the mechanical parameters of the wind turbines come with different frequencies and in a short while tend to eliminate each other (Akhmatov et al., 2003(a)). Therefore the resulting dynamic behaviour becomes smoother than in the case of a windfarm containing wind turbines with identical parameters.

This result is illustrated by a simulation example, where the fixed-speed (fixed-pitch) wind turbines in sections 1, 3, 6 and 8 have shaft stiffness K_S = 0.15 p.u./el.rad., and in sections 2, 4, 5 and 7 have K_S = 0.60 p.u./el.rad. The sections are located within the large offshore windfarm shown in **Figure 5.13**. The large offshore windfarm is compensated with a SVC unit of 50 MVAr, according to the data in **Table 5.1**. The simulated curves are plotted in **Figure 5.25**.

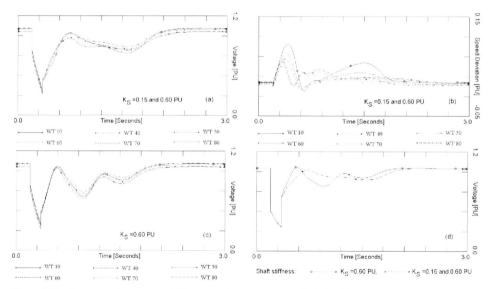

Figure 5.25   Comparison between a windfarm containing two groups of wind turbines with K_S =0.15 p.u./el.rad., respectively, K_S =0.60 p.u./el.rad. and a windfarm containing identical wind turbines with K_S =0.60 p.u./el.rad.: **(a)** – terminal voltage of selected wind turbines with the different shaft stiffness, **(b)** – generator rotor speed of selected wind turbines in the windfarm with different shaft stiffness, **(c)** – terminal voltage of selected wind turbines in the windfarm with identical parameters, K_S =0.60 PU/el.rad., **(d)** – voltage in connection point of the windfarm with different shaft stiffness compared to the case of the windfarm with identical shaft stiffness. Reprinted from Ref. (Akhmatov, 2003(b)), Copyright (2003), with permission of the copyright holder.

The use of the wind turbines with different shaft stiffness located within the large offshore windfarm may reduce grid voltage oscillations at the connection point of the windfarm as well as at the terminals of individual wind turbines.

Simulations are also made for a large offshore windfarm containing fixed-speed (fixed-pitch) wind turbines with different shaft stiffness K_S =0.15 p.u./el.rad., and K_S =0.30 p.u./el.rad. Finally, the simulations are repeated for cases where the fixed-speed wind turbines have fast-acting active-stall control during grid faults. In all these simulations, a reduction of the voltage oscillations when the wind turbines have different shaft stiffness is apparent. The use of fixed-speed wind turbines having different mechanical parameters may have such basic advantages as a reduction of voltage fluctuations in the power grid and better dampening of fluctuations in electrical and mechanical parameters of the wind turbines such as the speed, power and current.

5.7.7   Robustness of voltage control principles

In practical operations, control equipment failures may be caused by a severe event in the power grid. With regard to fixed-speed wind turbines, such control equipment failures may be experienced in dynamic reactive compensation units as well as in active-stall control systems of individual wind turbines. The control equipment must be robust, redundant and maintain uninterrupted operation when the power grid is subject to a short-circuit fault. Control equipment failures may not jeopardise short-term voltage stability and fault-ride-through capability of the wind turbines (Akhmatov et al., 2003(a)). The robustness of the control principles may relate to how large part of the control system may fail without loss of stability of the large offshore windfarm.

5.7.7.1   Failure of dynamic reactive compensation

When the wind turbines themselves have no control supporting grid-voltage re-establishment, maintaining of voltage stability will rely on the robustness of dynamic reactive compensation units applied together with the large offshore windfarm. To provide safety of operation, dynamic reactive compensation must have a number of reserve units. This may require that the installed reactive power capacity of the compensation unit is larger than used in investigations of short-term voltage stability (Akhmatov et al., 2003(a)) and that failure of parts of the equipment does not lead to the outage of the whole equipment (redundancy).

The reactive power capacity of the SVC unit in the range of 100 MVAr used in **Section 5.7.4.1**, is only the minimal demand to ride through the grid fault. This reactive power capacity does not account for the redundancy, e.g. the reserves which are necessary to compensate for possible equipment failure. If part of this 100 MVAr SVC unit failed and no reserves are used, this would presumably lead to voltage instability and protective disconnection of the large offshore windfarm.

Accounting for required reserves and redundancy of the equipment is responsibility of the manufacturer and commonly based on statistic methods.

5.7.7.2   Failure of active-stall control

In practical operations, the active-stall control systems of some of wind turbines may fail at a short-circuit fault in the power grid. This may lead to that these wind turbines do not reduce mechanical power at the short-circuit fault. More specifically, the active-stall control failure corresponds to that these wind turbines operate as fixed-pitch during the short-circuit fault (Akhmatov et al., 2003(a)).

As presented in Ref. (Akhmatov et al., 2003(a)), the large offshore windfarm will still maintain short-term voltage stability and ride through the grid fault when active-stall control fails in up to 25% of the active-stall controlled wind turbines of the windfarm. This result illustrates the robustness of the fast-acting active-stall control to stabilise the voltage at the grid faults, see **Figure 5.26**.

Case	Terminal voltage of selected wind turbines	Power supplied to grid from the windfarm
1		
2		
3		
4		
5		

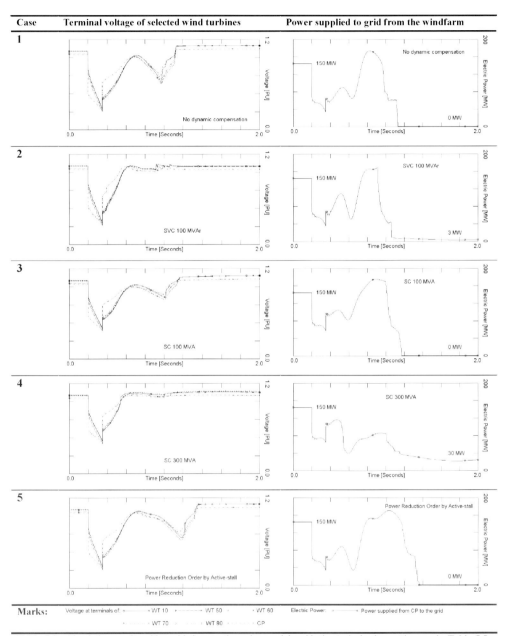

Marks: Voltage at terminals of: — WT 10 — WT 50 · WT 60 — WT 70 — WT 80 — CP   Electric Power: — Power supplied from CP to the grid

Table 5.3   Simulation results with enabled protective relay models with the typical relay settings as in **Table 5.5**. Power loss is present due to protective tripping of the wind turbines. Voltage recovers when the wind turbines have tripped. Reprinted from Ref. (Akhmatov, 2003(b)), Copyright (2003), with permission from the copyright holder.

A protective relay model is now a part of the simulation. The relay settings applied are the typical values from **Table 5.2**. The simulation results for the large offshore windfarm consisting of eighty wind turbines are collected in **Table 5.3** and represented with the use of the curves of the terminal voltage of selected wind turbines, voltage at the connection point of the windfarm and active power supplied from the windfarm to the power grid. The simulation results in **Table 5.3** are computed for different control arrangements applied to maintain short-term voltage stability of the power grid:

1) No control arrangements as in **Section 5.7.3**.
2) Incorporation of a 100 MVAr SVC unit as in **Section 5.7.4.1**.
3) Incorporation of a 100 MVA Synchronous Compensator with the excitation control of ASV1 as in **Section 5.7.4.2**.
4) Incorporation of a 300 MVA Synchronous Compensator with the excitation control of ASV3.
5) Power ramp using active-stall control as in **Section 5.7.5.5**.

The simulation results in **Table 5.3** show that there is no risk of voltage collapse in investigated cases. The main problem seems to be unnecessary disconnections of the wind turbines and a power loss. The analysis shows that the main reason of such disconnections may be that the relay settings, which are commonly applied in fixed-speed wind turbines and present in **Table 5.2** for over-current and under-voltage, are too conservative.

To avoid such unnecessary disconnections in this investigation, the relay setting for over-current is changed to 3.0 p.u. and its critical time is set to 20 ms. The relay settings for under-voltage are also changed to force the wind turbines to stay in operation. After the relay settings have been adjusted, the fixed-speed wind turbines within the large offshore windfarm ride through the applied short-circuit fault without disconnection and the grid voltage re-establishes. The simulation results become, then, similar to the results described in **Sections 5.7.3 – 5.7.4**.

In this investigation, adjusted relay settings are neither critical for wind turbine constructions nor for safely operation of induction generators and will not cause damage. Note that any changes in the protective relays must be coordinated with the wind turbine manufacturer.

To comply with the national Grid Codes and, in particular, with the requirement for fault-ride-through capability, a combined solution using dynamic reactive power compensation, active-stall control of the wind turbines acting at grid faults and also adjustment of relay settings can be necessary. Protective relay settings must be evaluated individually in each case of grid-connections of large windfarms. Investigations of short-term voltage stability using appropriate simulation tools and wind turbine models are useful and relevant to evaluate required relay settings of the wind turbines to be commissioned. With a disabled protective relay model, the simulated curves of the parameters monitored by the protective relay system, such as the terminal voltage, machine current, grid frequency and others, can provide relevant information for adjusting the protective relay settings in the wind turbines for gaining their fault-ride-through operation. Naturally, this requires a sufficient accuracy of the models applied in such investigations.

5.9 Summary

Voltage stability and fault-ride-through capability of fixed-speed wind turbines are closely related to the mechanism of over-speeding. The dynamic stability limit of such fixed-speed wind tur-

Figure 6.1   Part of a 242 × V47 windfarm commissioned at Woodward Mountain, Texas, USA, in 2001. Photo copyright Vestas Wind Systems. Reproduced with permission from Vestas Wind Systems.

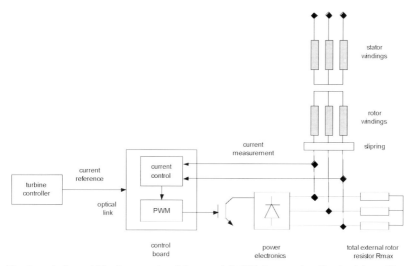

Figure 6.2   Layout of a variable-slip generator of the second OptiSlip® generation. Reprinted from (Vestas, 2003), Copyright (2003), with permission from Vestas Wind Systems.

6.1.2   Power and speed control

A part of the VMP controller is shown in **Figure 6.4**. When the power output of the grid connected variable-slip generator is below the rated power, the pitch angle is controlled in relation to the wind speed. In this operational situation, the pitch control is set to optimise the mechanical power of the rotor accordingly to the incoming wind. When this is part of the OptiSlip® wind turbine, this pitch control mode is termed OptiTip^{TM10}. The electrical power of the variable-slip generator is controlled by changing the power reference to the generator, which depends on the actual slip as shown in **Figure 6.5**.

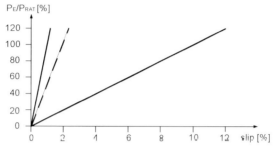

Figure 6.3 Comparison of power versus slip characteristics of a conventional induction generator with a 2% rated slip (dashed) and of a variable-slip generator with the OptiSlip®️ control (solid). Reprinted from (Vestas, 2003), Copyright (2003), with permission from Vestas Wind Systems.

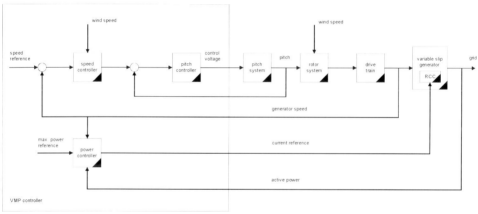

Figure 6.4 Control system. Reprinted from (Vestas, 2003), Copyright (2003), with permission from Vestas Wind Systems.

When wind speed is below the rated wind speed, the generator rotor slip is below 2%. Then, the power reference will follow the rated 2% slip characteristic plotted in **Figure 6.5**. The power controller is designed to smooth the active power by allowing the generator rotor slip to vary around the rated 2% slip characteristic within a given range. In this operational mode, the efficiency of the variable-slip generator will be equal to the conventional induction generator with 2% rated slip.

When the wind speed exceeds the rated wind, the generator rotor slip will be above 2% and the power reference is held constant at the rated power level. Dynamic variations in the wind speed will cause variations in the speed as the fast rotor current controller maintains the rotor current and the active power constant. This implies that the shaft torque will be almost constant reducing peak loads on the gearbox. In this operational mode, the speed controller will control the rotor speed by adjust-

[10] OptiTip is a trade mark of Vestas Wind Systems.

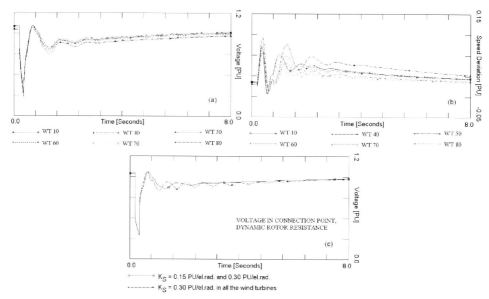

Figure 6.10 Wind turbines with different mechanical parameters and with the DRR control within the same windfarm:
(a) – terminal voltage of selected wind turbines, (b) – generator rotor speed and (c) – voltage in connection
point compared to the case of the windfarm equipped with identical wind turbines. Use of wind turbines
with different shaft stiffness within the large windfarm contributes to damping of voltage fluctuations at
grid faults. Reprinted from Ref. (Akhmatov, 2003(b)), Copyright (2003), with permission from the copyright holder.

Fluctuations of grid voltage and other parameters of the wind turbines are efficiently damped
when the wind turbines have different mechanical parameters (i.e. different shaft stiffness). This
result is similar to that reached for fixed-speed wind turbines equipped with conventional induction
generators and different shaft stiffness, see **Section 5.7.6**. The absence of mutual interaction between the doubly-outage induction generators with the DRR control is caused by the same reason as
in the case of the conventional induction generators (Akhmatov et al., 2003(a)), i.e. the absence of
synchronising torque in the case of induction generators. The use of the DRR control does not
change this fundamental behaviour of induction generators.

So far, this result is confirmed by practical experience from operational windfarms. Cases of mutual interaction between wind turbines with the DRR control have not yet been reported.

6.3   Reactive compensation

Doubly-outage induction generators with the DRR control are magnetised from the power grid in
the same way as conventional induction generators. In other words, the doubly-outage induction
generators absorb reactive power from the grid and then require static reactive compensation. This
is because the DRR control is not made to control reactive power, but only adjusts active power
according to the generator slip. Such partly variable-speed wind turbines equipped with doubly-
outage induction generators and the DRR control are either fully or no-load reactive compensated,
which depends on the Grid Code of the national system operator.

Dynamic reactive compensation, for example a SVC unit, may be required to improve short-term voltage stability and fault-ride-through capability of such wind turbines. The required capacity of dynamic reactive compensation units will, however, be reduced due to the use of the DRR control in combination with pitch control, when compared to fixed-speed wind turbines without such control.

6.4 Fault-ride-through capability

In the simulation example presented above, it is assumed that the wind turbines equipped with doubly-outage induction generators and the DRR control ride through grid faults without blocking the converter. This assumption may imply that the protective relay settings are tuned properly to avoid blocking of the converter and unnecessary disconnection of the wind turbines from the power grid.

6.4.1 Protective relay system

The protective relay system of the induction generator and the power electronics converter must be represented in investigations of power system stability. The protective relay system monitors the grid voltage, current in the stator and rotor, grid frequency, generator rotor speed and other parameters. When at least one of the monitored parameters exceeds its relay setting, the power electronics converter will be ordered to block, e.g. it stops switching and may trip. The converter blocking may lead to disconnection of the wind turbine. To make any conclusion about the action of the protective relay system of the power electronics converter, accurate models of the generator and the power electronics converter are required (Bolik, 2004). When the converter blocking is predicted using a generic, reduced model, the wind turbine manufacturer must be consulted to eliminate any doubts about the converter action predicted from such generic simulations.

With regard to the prediction of the converter blocking using a generic model, it is kept in mind that the IGBT switches must be protected against thermal and electrical overloads. Thermal overloads can be evaluated from the power losses in the rotor circuit which are given by $P_R = R_R(t) \cdot I_R^2$. In the simulation example presented, the power losses, P_R, are within an acceptable range. As illustrated in **Figure 6.11**, the power losses in the rotor circuit using the DRR control are between the power losses in the case of induction generators with a fixed rotor resistance (the conventional induction generators with short-circuited rotor circuits) of R_{R0} and $2 R_{R0}$.

Electrical overloads relate to excessive current transients in the rotor circuit. Such excessive current transients may damage the IGBT switches of the converter. Therefore the converter blocks almost immediately when the rotor current transients exceed a given limit (Akhmatov, 2002(b)). The simulated behaviour of the rotor current magnitude of selected wind turbines in the large windfarm is shown in **Figure 6.12**.

In the example presented, the rotor current transients reach the value of 2 p.u. which may cause blocking of the power electronics converter. The value reached by the rotor current transients during the grid voltage drop depends on the initial rotor current (before the grid fault has occurred). Therefore the power electronics converters of some wind turbines in the large windfarm may block when the power grid is subject to a short-circuit fault. This implies that the power electronics converter must be designed either to withstand such current transients or to block and then restart after a short time to ride through grid faults.

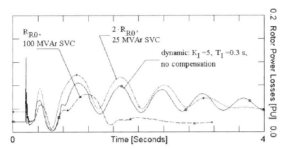

Figure 6.11   Evaluation of power losses in the rotor circuit, P_R, with the DRR control and the rotor circuit with a fixed rotor resistance. Reprinted from (Akhmatov et al., 2003(a)), Copyright (2003), with permission from Elsevier.

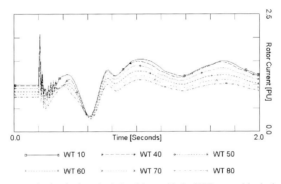

Figure 6.12   Rotor current magnitude of selected wind turbines with the DRR control in the large windfarm. Reprinted from (Akhmatov, 2003(b)), Copyright (2003), with permission from the copyright holder.

Representation of the protective relay system of the wind turbines is relevant in investigations of power system stability. Disconnection of wind turbines causes power loss in the grid and incompliance with the national Grid Codes. Therefore unnecessary disconnection of wind turbines must be avoided.

6.4.2   Converter blocking and restart

To improve the ride-through capability of wind turbines without over-sizing the IGBT switches of the converter, the converter may block at excessive current transients, but restart again when the grid operation returns to normal or when the current transients are eliminated (Akhmatov, 2003(d)).

When the power electronics converter blocks, the IGBT switches stop switching and the rotor circuit is closed through the total external rotor resistance as seen in **Figure 6.2**. The total external rotor resistance is the maximal external rotor resistance which can be gained by regular converter control, i.e. $R_{MAX} = R_{R0} K_{RMAX}$. The converter can be blocked so long as the power grid is in abnormal operation. In operation with the blocked converter, the input of the PI-controller of the regular converter control is kept at zero and the integrator output is set to K_{RMAX}. When the grid voltage recovers and the rotor current returns into the range of normal operation, the power electronics converter restarts, and, then, the input of the PI-controller receives the current error signal and regulates the dynamic signal $K_R(t)$. **Figure 6.13** shows a generic model for converter control with blocking and restart sequences of the power electronics converter. Notice that this converter action relates to the use of crow-bar protection when the total external rotor resistance, R_{MAX}, serves as the crow-bar.

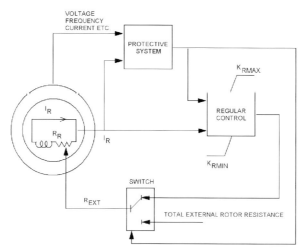

Figure 6.13   Use of crow-bar protection together with dynamic resistance control of doubly-outage induction generators.

6.5 Summary

Partly variable-speed wind turbines are equipped with doubly-outage induction generators with DRR and pitch control. The generators absorb reactive power from the grid and must therefore be compensated (static reactive compensation with the use of capacitor banks).

The active power and speed are controlled using the DRR control, which is arranged with using a power electronics converter connected to the rotor circuit. The total rotor resistance therefore consists of the static resistance of the rotor winding and the external rotor resistance dynamically adjusted by the power electronics converter. The external rotor resistance is varied from zero to the given upper limit, R_{MAX}, and controlled with regard to rotor current. This control is mainly used to improve power quality and reduce flicker emission from the wind turbines seen by the power grid.

This control may also improve short-term voltage stability and the ride-through capability of the wind turbines required by the national Grid Codes of many countries. The DDR control together with the pitch control prevent excessive over-speeding of wind turbines and, then, contribute to voltage re-establishment after a short-circuit fault in the power grid.

In investigations of voltage stability, the converter protection must be taken into account, because the converter may block due to excessive current transients in the rotor circuit initiated by a short-circuit fault. Without a fault-ride-through solution, the converter blocking may lead to disconnection of wind turbines and power loss in the grid.

To improve the fault-ride-through capability of the wind turbines, technical solutions with the converter blocking and restart can be applied. Such solutions can be arranged similarly to that with the crow-bar protection. Due to the sensitivity of power electronics converters to excessive current transients, the use of induction generator models with representation of the fundamental frequency (current) transients is preferred. When the converter blocking is predicted with the use of a generic control model, the wind turbine manufacturer must be consulted to eliminate any doubts about the action of power electronics converters and the fault-ride-through solution.

7   Variable-speed wind turbines with doubly-fed induction generators

As explained in **Section 3.1.3**, variable-speed operation provides better optimisation of power output produced by the wind turbine rotor. According to (Vestas, 2001), variable-speed operation makes it possible to increase the annual power production by approximately 5% when compared to fixed-speed operation. Variable-speed operation implies that the rotor speed, ω_M, may be in the range from –50% to +15% (its static limit) with respect to the synchronous speed depending on the incoming wind. Dynamically, the rotor speed may vary up to +30% over the synchronous speed (Vestas, 2001).

Variable-speed operation is gained using doubly-fed induction generators (DFIG) controlled by power electronics frequency converters. The DFIG is a wound rotor induction generator where the rotor circuit is connected to the power grid through the power electronics frequency converter on the slip rings of the rotor. The stator of the DFIG is directly connected to the power grid, see **Figure 7.1**. The power electronics frequency converter is controlled by switching the Insulated Gate Bipolar Transistor (IGBT) -switches (Heier, 1996; Müller et al., 2000).

The power electronics frequency converter is a back-to-back converter system consisting of two voltage-sourced converters (VSC) connected through a DC-link. The rotor circuit of the generator feeds into the rotor converter and operation of the rotor converter corresponds to adding an external voltage-phasor in series with the rotor circuit. This external voltage-phasor is controlled so that the electric frequency of the rotor circuit corresponds to the desired rotor speed. In normal power grid operations, rotor speed is adjusted by the rotor converter control in order to optimise power output which is why the rotor circuit operates at a variable electric frequency. The grid-side converter balances power injected into the DC-link of the back-to-back converter system versus active power exchanged with the power grid. In other words, the power electronics frequency converter provides a connection between the rotor circuit operating at variable electric frequency and the power grid being at a fixed grid frequency.

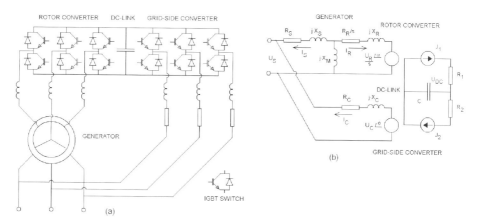

Figure 7.1   A doubly-fed induction generator with a power electronics frequency converter: **(a)** – schematic diagram, **(b)** – simplified representation for steady-state computations. Reprinted from Ref. (Akhmatov, 2003(e)), Copyright (2003), with permission from Multi-Science Publishing Company.

In steady-state operation, the total active power of the DFIG, P_E, is defined by the mechanical power of the rotor subject to incoming wind, P_M. Specifically, the total active power is the mechanical power of the rotor minus the losses in the mechanical construction and minus the electrical losses in the generator. The steady-state reactive power is initialised according to the control strategy chosen. Note that the active and reactive power are initialised independently characterising the control of such converter controlled generators and simplifies the initialisation process.

The following may be assumed with regard to the generator model initialisation.

1) The total active power at steady-state, P_E, is defined by the user in scenarios for the operational situations to be investigated with regard to short-term voltage stability of the grid. The total active power is the sum of the active power supplied from the stator terminals to the grid, P_S, and the active power exchanged at the rotor sliprings between the rotor circuit and the converter, P_R. The power losses in the converter are neglected, and so $P_E = P_S + P_R$.

2) The generator rotor slip, s, the wind speed and pitch angle may be found from the total active power as set in steady-state either with the use of an initialisation algorithm for the wind turbine rotor described in **Section 3.1** or by applying the steady-state wind versus power, wind versus speed and wind versus pitch-angle characteristics for the aerodynamic rotor received from the wind turbine manufacturer.

3) The reactive power exchanged between the DFIG system and the power grid is equal to the reactive power in stator, $Q_E = Q_S$, because the grid-side converter is considered to be reactive-neutral in normal operations of the power grid. This consideration is reasonable because the converter rating is around 25% of the generator rating and the grid-side converter is primarily applied to provide the active power exchange between the rotor circuit and power grid.

4) The reactive power in the stator will be zero in the case of a strong power system. The reactive power will also be set to zero when it focuses on optimisation of the power factor. In such situations, neither active control of the reactive power nor support of the grid voltage is required. In this case, the DFIG supplies only the active power to the grid and is completely excited by the rotor converter through the rotor circuit. The power factor of the DFIG system is then close to unity.

5) When support of the grid voltage is required in normal grid operations, the DFIG system will either supply or absorb an amount of reactive power. Then the steady-state reactive power is set so that the grid voltage is kept close to the desired value.

As can be seen, (i) the total active and the reactive power, P_E and $Q_E = Q_S$, (ii) the terminal voltage, U_S, and (iii) the generator rotor slip, s, of the DFIG system are known at the start of the initialisation. The initialisation of the generator model must result in (i) the power distribution between the stator and the rotor, P_S and P_R, (ii) the currents in stator and rotor, I_S and I_R, and (iii) the magnitude and the angle of the rotor voltage-source induced by the rotor converter, U_R. These parameters are applied to initialise the states of the generator model, which are the fluxes, shown in Eq.(4.4).

The frequency converter model will be initialised when knowing the rotor power, P_R, and the terminal voltage, U_S, and when neglecting losses.

7.1.1   Initialisation of generator model

Ref. (Akhmatov, 2002(a)) shows that the initialisation problem of the generator has been solved applying the super-imposing principle illustrated in **Figure 7.2**. Accordingly to this principle, the currents in the stator and in rotor are:

$$\begin{cases} I_S = I_{S1} + I_{S2}, \\ I_R = I_{R1} + I_{R2}, \end{cases} \tag{7.1}$$

have been induced by the respective voltage sources

$$\begin{cases} U_S = U_1, \\ U_R = U_2 / s. \end{cases} \tag{7.2}$$

Furthermore, the relations for the apparent power in the stator and in rotor:

$$\begin{cases} P_S + j \cdot Q_S = U_1 \cdot conj(I_S), \\ P_R + j \cdot Q_R = U_2 \cdot conj(I_R). \end{cases} \tag{7.3}$$

are applied together with Eqs(7.1) and Eq.(7.2).

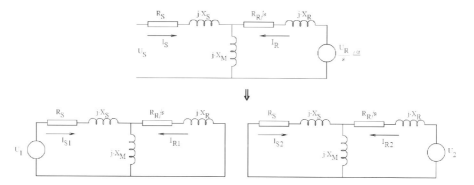

Figure 7.2   The super-imposing principle applied to solve the steady-state generator model. Reprinted from Ref. (Akhmatov, 2002(a)), Copyright (2002), with permission from Multi-Science Publishing Company.

The goal of this computation is to express the active and reactive power in the stator and rotor by the stator and rotor voltage, generator rotor slip and generator electrical parameters (data). The general expressions for the active and reactive power in the stator and rotor then become:

of the model as the power will be positive when produced and negative when absorbed by the generator.

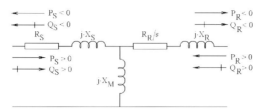

Figure 7.3   The relation between the sign convention of the power and the power-flow direction (supplied from or absorbed by the generator).

The sign inversion will imply that the power flow directions are opposite to those shown in **Figure 7.3**. In terms of the model equations, the sign inversion implies:

$$\begin{cases} P_S = -1 \cdot P_S, \ Q_S = -1 \cdot Q_S, \\ P_R = -1 \cdot P_R, \ Q_R = -1 \cdot Q_R. \end{cases} \tag{7.9}$$

When the inverse sign convention is required for further computations, Eqs.(7.9) may be applied to the results of the iterative routine using Eqs.(7.1-7.8).

7.1.3   A generator model initialisation example

In this example, a 2 MW DFIG is in rated operation, $P_E = 1.0$ p.u. (with the inversed sign convention applied). The corresponding generator rotor slip is $s = -0.10$. The stator voltage, U_S, is 1.0 p.u. and the reactive power control is set to optimise the power factor to unity, $Q_s = 0$. The generator parameters are $R_s = 0.00779$ p.u., $X_s = 0.07937$ p.u., $X_M = 4.1039$ p.u., $R_R = 0.0082$ p.u. and $X_R = 0.40$ p.u., see Ref. (Ledesma et al., 1999). The solution for the active and reactive power of the DFIG is computed using Eqs.(7.4) (with the inversed sign convention) and plotted in **Figure 7.4** as functions of the phase difference, α, with the rotor voltage magnitude, U_R, as a parameter. The general solution for the active power of the wind turbine driven generator must be in the range 0.0 to 1.0 p.u. The active power supplied to the grid must not be negative as the negative power would correspond to the motor operation (with the use of the inversed sign convention). The generator cannot supply more power than the mechanical power produced by the wind turbine rotor which is the active power does not exceed 1.0 p.u. in normal operation.

In general, the solution for the reactive power might be in the range -1.0 p.u. to 1.0 p.u. restricted by the generator power rating. Normally the reactive power range is also restricted by the power rating of the power electronics converter, as the converter excites the generator through the rotor circuit and controls the reactive power of the generator. These limits are manufacturer-specific and may also depend on the active power supplied by the generator (Bolik, 2003).

In this example, the solution related to $P_E = 1.0$ p.u. and $Q_s = 0.0$ p.u. are of interest, which defines the intervals of interest for the phase difference, α, and for the rotor voltage magnitude, U_R. By inspection of the curves plotted in **Figure 7.4**, the intervals of interest are from 200° to 210° for

α and from 0.11 p.u. to 0.12 p.u. for U_R, respectively. The curves for the active power, reactive power and current in the stator and rotor within the intervals of interest are plotted in **Figure 7.5**.

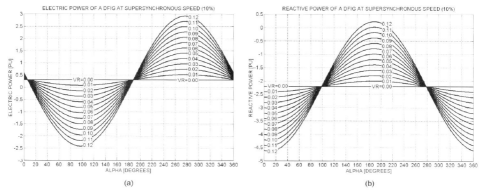

Figure 7.4   Computed solutions for doubly-fed induction generator: (a) - the electric power, P_E, and (b) the reactive power, Q_S, at the given terminal voltage and given slip plotted as functions of the phase difference, α, and rotor voltage magnitude, U_R. Reprinted from Ref. (Akhmatov, 2002(a)). Copyright (2002), with permission from Multi-Science Publishing Company.

Therefore, the final solution for the initialisation of the DFIG can be found within the given (reduced) intervals of interest for the phase difference, α, and for the rotor voltage magnitude, U_R. Applying iterations within the intervals of interest, the final solution is found at $\alpha =206°$ and $U_R =0.115$. The solution for the active and reactive power of the DFIG is sensitive with regard to the phase difference, α, and also the rotor voltage magnitude, U_R.

Furthermore, as found in this example,

$$\begin{cases} P_S \approx (1 \mp s) \cdot P_E, \\ P_R \approx \pm s \cdot P_E, \\ U_R \approx \pm s \cdot U_S. \end{cases} \qquad (7.10)$$

These relations are said to be general for the DFIG in the case when $Q_S =0$ (Cadirci and Ermis, 1992; Pena et al., 1996). During rated operations and $Q_S =0$, the current in the stator and rotor is close to 1.0 p.u. The general rule for the power distribution in the DFIG is that the active power is always supplied from the stator terminals to the power grid, independently of the value of the generator rotor slip (Heier, 1996; Zhang et al., 1997; Cadirci and Ermis, 1992). In contrast, active power is supplied from the rotor circuit to the power grid at super-synchronous operation and absorbed by the rotor circuit from the power grid at sub-synchronous operation. When the rotor circuit absorbs the active power from the grid, the active power supply from the stator mains to the grid is increased by the same value to balance the rotor mechanical power. **Figure 2.11** presents the power distribution in the DFIG in different operational modes.

In synchronous operation, the generator rotor slip is zero, which means that the generator rotor speed corresponds to the electric system speed (the mechanical rotor frequency multiplied by the number of pole-pairs and the grid frequency are all equal) and the frequency of the field induced in the rotor circuit is zero. In synchronous operation, the active power is absorbed by the rotor circuit by means of covering the resistive losses in the rotor circuit and of magnetising the generator. In synchronous operation, the reactive power of the rotor circuit is zero, as indicated by Eqs.(7.7).

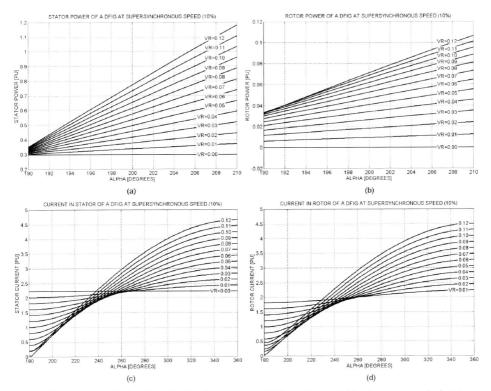

Figure 7.5   Computed solutions for: **(a)** - electric power supplied from stator and **(b)** - from rotor terminals, **(c)** - current in stator and **(d)** - in rotor within the reduced intervals for the phase difference, α, and for the rotor voltage, V_R around the interval of interest. Reprinted from Ref. (Akhmatov, 2002(a)), Copyright (2002), with permission from Multi-Science Publishing Company.

General statements by Eqs.(7.10) are applied for simple and fast verification whether the found solution for initialisation of the DFIG is in agreement with expectations for the power flow in the stator and rotor.

7.1.4   Initialisation of converter model

When the rotor power, P_R, is found from the initialisation routine of the DFIG, the charging current of the DC-link can be initialised as:

$$J_1 = \frac{P_R}{U_{DC}},\qquad\qquad(7.11)$$

and the discharging current of the dc-link will be initialised to balance the charging current:

$$
\begin{cases}
J_2 = J_1, \\
J_2 = \dfrac{P_{GC}}{U_{DC}}.
\end{cases}
\tag{7.12}
$$

The electric power P_{GC} denotes the active power transferred through the DC-link from the rotor circuit to the grid-side converter, and neglected losses, $P_{GC} = P_R$. The expressions given by Eq.(7.11) and Eq.(7.12) are found by inspection of the converter representation given in **Figure 7.1.b**. The relations for the converter model initialisation with power losses in the DC-link may be found in (Akhmatov, 2003(e)).

The DC-link voltage in steady-state operation, U_{DC}, is coupled to the AC- voltage of the grid at the grid-side converter mains, U_S.

$$
U_S = \frac{\sqrt{3}}{2 \cdot \sqrt{2}} \cdot U_{DC} \cdot p_M,
\tag{7.13}
$$

where p_M denotes the modulation depth of the grid-side converter and U_S and U_{DC} are measured in kV. When assuming $p_M \leq 1$, then the initial DC-link voltage becomes:

$$
U_S \leq \frac{\sqrt{3}}{2 \cdot \sqrt{2}} \cdot U_{DC} \Rightarrow U_{DC} \geq \frac{U_S}{0.6124}.
\tag{7.14}
$$

The active power at the grid-side converter terminals, P_C, is initialised knowing the total active power supplied by the DFIG to the grid, P_E, and the active power supplied from the generator stator terminals, P_S.

$$
P_C = P_E - P_S.
\tag{7.15}
$$

When the power losses in the grid-side converter terminals are disregarded, the active power at the grid-side converter terminals becomes:

$$
P_C = P_R.
\tag{7.16}
$$

The reactive power at the grid-side converter terminals, denoted by Q_C, is assumed to be zero (the grid-side converter is reactive neutral with the power grid in normal operation). Hence, the expressions describing the initialisation of the grid-side converter are:

$$
\begin{cases}
P_C + j \cdot Q_C = U_S \cdot conj(I_C), \\
I_C = \dfrac{U_C - U_S}{R_C + j \cdot X_C}.
\end{cases}
\tag{7.17}
$$

The goal is to find the magnitude and phase angle of the grid-side converter voltage source, U_C. When the power losses in the converter mains are disregarded, this is done by iteration of Eq.(7.17).

Power losses in the grid-side converter mains, ΔP_C, are caused by resistance of the smoothing inductor, i.e.:

$$\Delta P_C = R_C \cdot | I_C |^2. \tag{7.18}$$

Disregarding such power losses relates to neglecting the resistance of the smoothing inductor, R_C. However, the resistance of the smoothing inductor cannot always be disregarded due to simplification of the model because this resistance can be important for dampening current transients in the grid-side converter mains at grid disturbances (Svensson, 1998).

Presence of the resistive component R_C in the model leads to power losses in the smoothing inductors of the grid-side converter, which is why P_R is not equal to P_C. Then, the initialisation routine of the DFIG system model must contain a number of iterative loops with the power balance $P_E = P_S + P_C$ as the checking condition (Akhmatov, 2003(e)).

7.1.5 Validation of the DFIG system model initialisation

A precise validation of the DFIG model initialisation must be based on measurements or comparison of computation results from a reliable simulation program. For instance, the simulation program applied by the wind turbine manufacturer can be very useful, because such programs are always verified.

Validation of the steady-state operation of the DFIG is made using simulation results and specific generator parameters offered by the manufacturer Vestas Wind Systems for a 1.75 MW OptiSpeed[®] wind turbine generator (Akhmatov, 2003(b)). In validation of the initialisation routine presented in prior sections, it is assumed that the grid voltage is fixed $U_s = 1.0$ p.u. (690 Volts). Furthermore, power losses in the converter and slip-rings are disregarded. However, such power losses are included in the computations using the model of Vestas Wind Systems. Presumably, this may lead to some discrepancies when the results of the DFIG model initialisation are compared to the results reached by the model of Vestas Wind Systems.

For each single case to be validated, the manufacturer Vestas Wind Systems have given the total active power, the power distribution in the stator and rotor and the current in the stator and rotor. The generator rotor slip and the reactive power $Q_s = 0$ are also given by Vestas Wind Systems for each validation case.

Applying the total active power, P_E, the reactive power, Q_S, the slip, s, and the voltage in the stator, U_S, the initialisation routine is used to define the power distribution in the stator, P_S, and the rotor, P_R, and the current in the stator, I_S, and the rotor, I_R. The results of the initialisation routine found for each case are compared to the results offered by the manufacturer. **Table 7.1** presents some of validation results as deviations between the results given by the manufacturer and those found by the initialisation routine:

$$\Delta X = 2 \cdot \left| \frac{X_1 - X_2}{X_1 + X_2} \right| \cdot 100\%, \tag{7.19}$$

where ΔX denotes the deviations given in **Table 7.1**, X_1 and X_2 are the results given by Vestas Wind Systems and the results of the initialisation routine, respectively. As can be seen, the results reached with the initialisation routine are in good agreement with those offered by Vestas Wind Systems. For the values of the stator and rotor current, small discrepancies are however seen.

In computations with the initialisation routine of the DFIG model, the value of the terminal voltage is consequently kept at the rated value of U_s =690 Volts. In the computations by Vestas Wind Systems, the terminal voltage has been approximately its rated value. In the initialisation routine of the DFIG model, the power losses in the converter are disregarded, but the converter losses are represented in the computations by the manufacturer as in a physical DFIG system. This then explains possible discrepancies in the current seen in the validation.

When the DFIG is close to the synchronous speed, the power exchanged between the rotor circuit and power grid through the power electronics converter is small, which is why even small discrepancies can lead to a large relative discrepancy (in %). This explains the 25.8% discrepancy in the rotor power in the first row of **Table 7.1**.

With regard to the generator model, the results presented in **Table 7.1** are for the super-synchronous operation only and validate the expressions of Eq.(7.4) for the active power in the stator and rotor and the expressions of Eq.(7.6) for the stator and rotor current. The model of the DFIG system must also predict accurate results at synchronous operation, e.g. when the generator rotor slip is zero s =0, due to possible risk of divergence of the generator model. Therefore, the manufacturer Vestas Wind Systems suggested more validation and offered the results of their model obtained for the synchronous operation. The results of this validation correspond to the validation of Eq.(7.7) for the active power in the stator and rotor and the expressions of Eq.(7.6) for the current in the stator and rotor, see **Table 7.2**.

Total power, p.u.	Slip, -	Stator power, %	Rotor power, %	Stator current, %	Rotor current, %
0.29	-0.0133	0.29	25.8	0.23	1.68
0.57	-0.1200	0.25	2.22	0.26	3.18
0.85	-0.1200	0.29	2.62	0.17	3.74
1.00	-0.1200	0.35	3.13	0.29	2.70

Table. 7.1 Validation of the initialisation routine of the DFIG model using computations given by the manufacturer Vestas Wind Systems. Reprinted from Ref. (Akhmatov, 2002(a)), Copyright (2002), with permission from Multi-Science Publishing Company.

Total power, PU	Slip, -	Stator power, %	Rotor power, %	Stator current, %	Rotor current, %
0.29	0.0	0.02 %	6.45 %	0.72 %	1.29 %
0.57	0.0	0.08 %	10.25 %	1.32 %	1.67 %
0.85	0.0	0.06 %	7.90 %	2.84 %	2.23 %
1.00	0.0	0.10 %	6.83 %	3.19 %	1.90 %

Table 7.2 Verification of initialisation of the DFIG model against computations made by the manufacturer Vestas Wind Systems at synchronous operation. Reprinted from Ref. (Akhmatov, 2003(b)), Copyright (2003), with permission from the copyright holder.

cost of the converters (Akhmatov, 2002(a)). However the use of smaller power electronics converters introduces restrictions on the current transients and makes the power electronics converters to be among the most sensitive parts of the variable-speed wind turbines with regard to the grid voltage drop.

To protect the IGBT-switches from overloads and risk of damage, the current in the rotor circuit and the grid-side converter mains must be limited (Akhmatov, 2002(a)). Therefore, the converter protective system monitors several electrical parameters such as the current in the stator and in the rotor circuit of the generator and in the grid-side converter mains, the DC-link voltage, the grid voltage, the electrical frequency, etc. When at least one of the monitored parameters exceeds its respective relay settings, the protective system orders the power electronics converter to block. When the converter blocks, it stops switching and may trip. This blocking leads to loss of control of the active and reactive power and may even lead to disconnection of the wind turbine itself.

As the power electronics converters are mostly blocked due to excessive current transients (Akhmatov, 2002(b)), such transients must be represented in the DFIG system model applied in investigations of short-term voltage stability. Thus, the transient, fifth-order model is preferred in such investigations because this model will predict more accurate behaviour of the current transients and also of the converter action at the grid voltage drops. The reduced, third-order model disregards the fundamental-frequency transients in the machine current and then cannot always represent the converter action at the grid voltage drops. The reduced, third-order model should not be used to evaluate the fault-ride-through capability of the DFIG system due to the above reasons.

As demonstrated in (Akhmatov, 2002(b)), power electronics converters of the DFIG systems may already block when the grid voltage drops abruptly from 1.0 p.u. to 0.8 p.u. This result illustrates how sensitive the power electronics converters can be regarding such grid voltage drops. **Figure 7.6** compares the simulated response of the converter controlled DFIG system modelled with the use of the transient, fifth-order model versus the reduced, third-order model of the generator to the short-circuit fault resulting in a voltage drop to 0.8 p.u. These results must be seen together with the description of the converter protection given in **Section 7.7**. The simulations show that the transient, fifth-order model predicts the rotor converter blocking due to excessive transients in the rotor circuit whereas the reduced, third-order model predicts that the converter system rides through this grid fault. This result shows that the use of the transient, fifth-order model of the DFIG is required in investigations when the voltage drop at the generator terminals is already around 20% which corresponds to the case of a single-phase short-circuit fault, **Section 4.3.1**.

When the transient, fifth-order model of the DFIG may predict the converter blocking and probable disconnection of the wind turbine at such a voltage drop, the reduced, third-order model of the DFIG may predict uninterrupted operation of the converter and its control in the same operational situation. This discrepancy is model-dependent and can be significant for the outcome of investigations of short-term voltage stability.

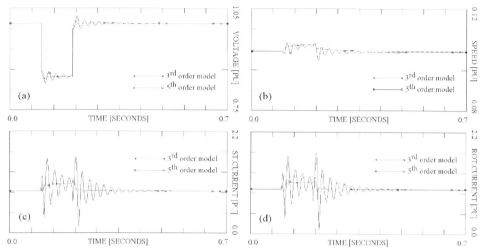

Figure 7.6 Illustration of the model-dependent difference in response of the converter controlled DFIG system to a short-circuit fault: (a) - terminal voltage, (b) - generator rotor speed, (c) - stator current and (d) - rotor current. Reprinted from Ref. (Akhmatov, 2002(a)), Copyright (2003), with permission from Multi-Science Publishing Company.

During the work of (Akhmatov, 2003(b)), the contact was taken to the manufacturer Vestas Wind Systems producing OptiSpeed® variable-speed wind turbines with DFIG to clarify the generator model details for investigations of short-term voltage stability. The manufacturer Vestas Wind Systems agreed that it can be necessary to apply a transient, (at least) fifth-order model of the DFIG to predict the exact converter action at voltage drops (Akhmatov, 2002(b)).

In Ref. (Pöller, 2003), the response of the transient, fifth-order model was compared to the response of the reduced, third-order model of the DFIG when the voltage at the DFIG terminals dropped to zero. In this case, both generator models predicted the same result - the power frequency converter blocked when the grid fault occurred and restarted when the grid fault was removed. Note that this model comparison does not prove anything regarding the generator modelling as this investigation was carried out for the such an efficient voltage drop at which the power electronics converter would block regardless of what kind of generator model was applied.

Although the reduced, third-order model of the DFIG with neglected fundamental-frequency transients in the machine current seems to be over-simplified, this model still can be applied in stability investigations with certain assumptions. Such assumptions can be that the rotor converter always blocks when the terminal generator voltage drops below a given setting, for example 0.8 p.u., for longer than a given period, for example 20 ms, which is the inverse of 50 Hz. The values of such settings and periods can be received from wind turbine manufacturers.

This assumption is useful because many simulation tools only have reduced, third-order models of induction generators. However, this assumption does not necessarily represent a real situation which is why the obtained results must be treated with percussions and discussed with the manufacturers. In this book, all the simulation cases are run with the use of the transient, fifth-order model to eliminate doubts about the converter operations. The discussion about the converter protection by excessive current transients will be continued in **Section 7.7**.

The control of the DC-link voltage, U_{DC}, is arranged with the two PI-controllers in series and the control of the reactive current, i_β, by a single PI-controller (Schauder and Mehta, 1993). The DC-link voltage is balanced by varying the active-current reference, $i_{\alpha,REF}$, in response to the DC-link voltage error. This controls the active power exchange between the grid-side converter and the power grid. The reactive power is regulated according to the setting of the reactive-current reference, $i_{\beta,REF}$. By default the reactive-current reference is set to zero, which is why the grid-side converter operates reactive-neutral. The output of the grid-side converter control is the (α,β)-components of the voltage induced by the grid-side converter $U_C = [u_{C\alpha}, u_{C\beta}]^T$. The switching dynamics of the IGBT-switches of the grid-side converter are neglected as the grid-side converter is able to follow the reference values $u_{C\alpha}$ and $u_{C\beta}$ at any time.

7.4.3.1 Optional cross-coupling

In some control systems of the grid-side converters, voltage-compensation using cross-coupling is applied. The cross-coupling in terms of $(U_S - X_C i_\beta)$ and $(X_C i_\alpha)$ is shown by dotted lines in **Figure 7.8.d**. The use of such dotted lines must indicate that this part of the grid-side converter control is optional and not used in every application.

7.4.3.2 Optional control of reactive power and voltage support

It is feasible to set the grid-side converter to control the reactive power and to support the voltage in the vicinity of the connection point. This control may be arranged in several ways by addition of the control block setting the reactive power set point of the grid-side converter according to the grid voltage deviation. **Figure 7.8.e** presents the control system consisting of two PI-controllers in series producing the reference signal of the reactive current, $i_{\beta,REF}$, from the voltage error signal. This control system is applied together with the conventional control of the grid-side converter given in **Figure 7.8.d**.

When the grid-side converter control is applied to control the reactive power and to support the grid voltage, this will require controlled co-ordination between the rotor converter and the grid-side converter as the rotor converter is already used to excite the generator (Akhmatov, 2003(e)). The reactive power control using a grid-side converter can be relevant when the power grid is subject to a short-circuit fault and the rotor converter has blocked. The reactive power control provided by the grid-side converter may contribute to faster re-establishment of the grid voltage in operation situations when the rotor converter has blocked (Akhmatov, 2002(b)).

When variable-speed wind turbines are commissioned in large (offshore) windfarms and subject to the national Grid Codes, the wind turbines can be obliged to provide ancillary system services. In this case, it is thinkable that the external system accesses the reactive power control of the grid-side converters and orders an amount of reactive power to the grid. As the power rating of the grid-side converters is small compared to that of the generators, the reactive power which is exchanged by the grid-side converters and the power grid is limited. The link between the external system and this optional control of the grid-side converter is illustrated in **Figure 3.7**.

7.4.3.3 Reduced converter model

In investigations of short-term voltage stability, representation of the DC-link and the grid-side converter control is often disregarded (Akhmatov, 2002(b); Pena et al., 1996; Gjengedal et al., 1999; Pena et al., 2000; Røstøen et al., 2002). This simplification is made due to two reasons (Stapleton et al., 2002), i.e. the:

1) Power rating of the converter is small compared to the generator. Then, the impact of the grid-side converter on short-term voltage stability is small compared the impact of the generator.
2) Control of the grid-side converter is fast. Then, it is assumed that the grid-side converter is able to keep the DC-link voltage at the desired value at any time.

When the above simplifications are applied, the power electronics converter is only represented with the rotor converter and its control, termed a reduced converter model. When such a reduced converter model is applied, the DC-link voltage is assumed to be stationary in any operational situations of the power grid. The active power supplied by the grid-side converter mains to the power grid, P_C, is equal to the rotor power, P_R, at any time. The reactive power of the grid-side converter is also zero at any time, $Q_C =0$. In this case, the rotor converter control is represented as shown in **Figure 7.8.a-c** and the model of the grid-side converter and its control is shown in **Figure 7.9**. The use of a reduced converter model may reduce the complexity of the converter representation and control and speed up for simulations.

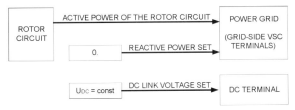

Figure 7.9   Representation of the DC-link and the grid-side converter in a reduced converter model.

Application of the reduced converter model may however introduce inaccuracy in the converter response to grid faults. As found in (Akhmatov, 2003(b)), application of the reduced converter model may lead to:

1) In a physical back-to-back converter system, fluctuations of the DC-link voltage may occur when the grid voltage drops at the grid fault. The DC-link voltage is one of the parameters monitored by the converter protection system. When fluctuations of the DC-link voltage are excessively intense, the converter may block. This behaviour will not be represented in a reduced converter model.
2) When the reduced converter model is applied, the grid-side converter cannot be set to dynamic control of the reactive power and support the grid voltage.
3) The use of the reduced converter model may influence the dampening characteristic of the converter control model with regard to attenuation of the shaft torsion oscillations. In other

words, the reduced model of the power electronics converter may predict poorer dampening of the shaft torsion oscillations than will be achieved by the physical converter, see **Section 7.8.3**.

These model-dependent discrepancies are not trivial as they may influence results of dynamic simulations. Being closer to the physical approach, the detailed converter model predicts more accurate results than the reduced model. The use of the detailed model is therefore recommended in investigations of short-term voltage stability.

7.5 Interface to the power grid model

The dynamic model of the DFIG system must be interfaced with the power grid model to exchange results of simulations for any time step. Viewed from the power grid model, the DFIG system may be represented by two Norton-equivalents (Akhmatov, 2002(a)). This representation contains the Norton-equivalent of the generator with the transient impedance $Z'=R_S+jX'$ and the current source, J_G, as described in (Kundur, 1994; Feijóo et al., 2000) and the Norton-equivalent of the grid-side converter with the impedance of the converter mains, $Z_C=R_C+jX_C$, and the current source, J_C, as suggested in (Akhmatov, 2002(a)). The DFIG system interface is shown in **Figure 7.10.a**. When superimposing these two Norton-equivalents, the resulting interface becomes as shown in **Figure 7.10.b**.

The simulation tools applied for investigations of short-term voltage stability often disregard the fundamental-frequency transients in the power grid models. In this way, the network solution is the load-flow solution for the power grid computed at each time step. When interfacing the transient, fifth-order model of the DFIG to such simulation tools, it is necessary to acknowledge the constraint of the dynamic simulation tools. The stator flux transients are therefore represented internally in the DFIG model, which is sufficient for representing the fundamental-frequency transients in the generator current in such stability investigations (Akhmatov and Knudsen, 1999).

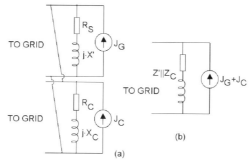

Figure 7.10   The Norton-equivalent of the DFIG system: (a) - representation using two Norton-equivalents and (b) - representation with superimposed Norton-equivalent. Reprinted from Ref. (Akhmatov, 2002(a)), Copyright 2002, with permission from Multi-Science Publishing Company.

7.6 Numeric stability of models

The model is numerically stable when it predicts the same results within an interval of the fixed integration steps when the simulation tool is a fixed-step tool, or within an interval of the maximal integration steps when the simulation tool is a variable-step tool. Furthermore, the model must neither constrain the other (standardised) models nor the network solver of the simulation tool.

Figure 7.11 illustrates the characteristic time constants of different partial models and physical processes in the variable-speed wind turbines equipped with DFIG and power electronics converters. These characteristic time constants are in the range from parts of millisecond to tens of seconds (Akhmatov, 2003(e)). Such a difference in the characteristic time constants may introduce numeric stability issues when applying the model in a simulation tool with a fixed integration step, or when a relatively large maximal (variable) step is required for stability investigations using a variable-step tool.

The numeric stability of the model can sufficiently be improved with the use of the so-called stiff integration routines applying a smaller (variable) time step within the model (Ibrahim, 1997). When the model is applied in the simulation tool with a fixed time step, the stiff integration routine must be implemented and applied internally in the model whereas the network solver is still based on the fixed time step algorithm.

In ref. (Akhmatov, 2003(e)) the numeric stability of a variable-speed wind turbine model using such a stiff integration routine is evaluated. The evaluated user-written model is implemented in the fixed-step simulation tool PSS/E^{TM}. For this evaluation, the protective relay model was disabled to investigate the numeric response of the variable-speed wind turbine model to a large disturbance. The power electronics converter control was represented by all the control systems shown in **Figure 7.8**. The simulations were executed for the case of a 3-phase short-circuit fault in the vicinity of the generator terminals. At the grid fault, the generator terminal voltage dropped down to 0.15 p.u.

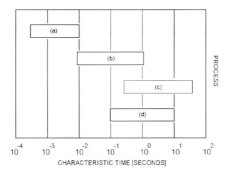

Figure 7.11   Processes versus their respective characteristic time constants: **(a)** – interaction between the converters and the generator, **(b)** – fundamental-frequency model of the generator, **(c)** – pitch control, wind turbine mechanics, **(d)** – interaction between grid and wind turbine generator at balanced faults. Reprinted from Ref. (Akhmatov, 2003(e)), Copyright (2003), with permission from Multi-Science Publishing Company.

The simulations were carried out with different fixed integration time steps in the range 0.1 ms to 10 ms. The simulation results for the integration time steps up to 4 ms are plotted in **Figure 7.12**. The results are found to be identical for the fixed integration time step of 0.1 ms to 2 ms. At the time step of 4 ms, a slight deviation in the simulation results was seen compared to the results reached with smaller integration time steps. When applying larger integration time steps, the deviation increases.

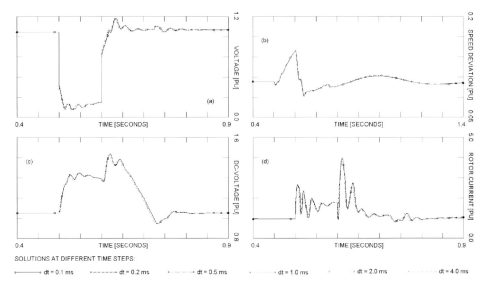

Figure 7.12   Illustration of numeric stability of variable-speed wind turbine model. The model resulting parameters computed in the simulation tool PSS/E^{TM} with different fixed integration time steps: **(a)** – terminal voltage, **(b)** – generator rotor speed, **(c)** – DC-link voltage, **(d)** – rotor current. Protective relay model is disabled in simulations. Reprinted from Ref. (Akhmatov, 2003(e)), Copyright 2003, with permission of Multi-Science Publishing Company.

Use of a smaller than 0.1 ms integration step was not reasonable due to a risk that slower processes in the present model and in the standardised models of the simulation tool would accumulate computational errors and slow computations. Note that such standardised models of the simulation tool PSS/E^{TM} are normally executed at the default integration time step of 10 ms.

7.7   Protective system of power electronics converters

When the power grid is subject to a short-circuit fault, the grid voltage drops and introduces transients in the rotor circuit of the DFIG. If the rotor current transients become excessive, the power electronics converter may block. Operation of the variable-speed wind turbine will then depend on the converter action and the protection applied for the converter (Akhmatov, 2003(d)).

When the national Grid Code does not require fault-ride-through operation, the power electronics converters of variable-speed wind turbines may block and the wind turbines may disconnect from the grid. The local wind turbines commissioned in Denmark before July 2004 are not subject to such a requirement, which is why the variable-speed wind turbines equipped with DFIG and power electronics converters may trip. Subsequent disconnection of wind turbines will lead to power loss which may require incorporation of power reserves to balance the active power and to control the grid frequency. Therefore the protective system of power electronics converters must be represented in the dynamic model of such variable-speed wind turbines to be applied for investigations of short-term voltage stability.

Simulation results shown in **Figure 7.12** indicate that the power electronics converters may block due to the following reasons.

1) Excessive over-voltage in the DC-link as shown in **Figure 7.12.c**.
2) Excessive over-current in the rotor circuit, see **Figure 7.12.d**.
3) Excessive over-current in the grid-side converter mains (not shown in **Figure 7.12**, but notified in this study case).

As found in (Akhmatov, 2002(b)), the most common reason for the rotor converter blocking is the protection of the IGBT-switches from excessive current transients in the rotor circuit. Such excessive current transients may occur due to the following events (Akhmatov, 2002(b)).

1) When the power grid is subject to a short-circuit fault.
2) At the moment when the fault has cleared.
3) At the moment of the generator re-connection to the grid or of the converter re-start (Akhmatov, 2003(d)).

As the reduced, third-order model does not compute current transients, the protection system can be implemented, as suggested in **Section 7.2.1**.

Note that the results shown in **Figure 7.12** are simulated on the assumption that the wind turbine is at rated operation (in strong wind) and that the short-circuit fault causes a significant voltage drop in the vicinity of the wind turbine terminals. In its rated operation, the initial current of the generator and the power frequency converter is closer to the protective relay settings. A significant voltage drop will excite larger current transients in the rotor circuit. Therefore the rated operation of the wind turbines and the occurrence of the short-circuit fault with a significant voltage drop correspond to the worst case with respect to maintaining uninterrupted operation of the rotor converter.

For this variable-speed wind turbine concept, concerns about the protection of the rotor converter, the grid-side converter and the DC-link from excessive over-current in the rotor converter (and also in the grid-side converter mains) and excessive over-voltage in the DC-link have a general character. This result is important for understanding the interaction between the DFIG-based variable-speed wind turbines and power grids. Accurate representation of the protective system is important for demonstration of the fault-ride-through capability of the variable-speed wind turbines equipped with DFIG and power electronics converters.

7.7.1 Converter blocking

Power electronics converters of variable-speed wind turbines must be modelled with representation of their protective systems. The protective systems monitor the operation of the converter, the generator and the power grid (at the generator terminals, for example) and may order the converter to block when abnormal operation is registered. Converter blocking must protect the IGBT-switches from electrical and thermal over-loads (Akhmatov, 2002(b)). As usual, the protective system monitors (i) the rotor current, (ii) the current in the grid-side converter mains, (iii) the DC-link voltage, (iv) the grid voltage, (v) the grid frequency, etc. and orders the converter to block when at least one of the monitored parameters exceeds its respective relay setting. Typical characteristic time of the converter blocking can be faster than few ms (Akhmatov, 2002(b)).

Since the power rating of the rotor converter is limited, the upper limit of the rotor current passing the IGBT-switches of the rotor converter is not much larger than the rated rotor current. As found in (Akhmatov, 2002(b); Akhmatov, 2003(d)), rotor converter blocking is mostly caused by excessive current transients in the rotor circuit. Blocking of the rotor converter can be arranged in two different ways such as (i) opening the rotor circuit for fast demagnetising of the generator, (ii) closing the rotor circuit through a crowbar.

7.7.1.1   Rotor circuit "opens"

When the rotor converter is ordered to block, the IGBT-switches of the rotor converter stop switching and open. When the IGBT-switches have opened, the rotor circuit "opens" and looks into the DC-link capacitor through the diode-bridge, see **Figure 7.13**. When the voltage magnitude of the AC rotor voltage on the rotor circuit terminals is larger than the DC-voltage across the capacitor, the diodes are leading and the current is going from the rotor circuit into the DC-link. The voltage induced at the rotor terminals decreases whereas the DC-voltage across the capacitor may increase. When the capacitor has charged to a DC-voltage that is larger than the magnitude of the AC rotor voltage, the diodes block and the current flow in the rotor circuit stops. Note that the grid-side converter may maintain uninterrupted operation to balance the DC-link voltage, see also **Section 7.8.2.1**.

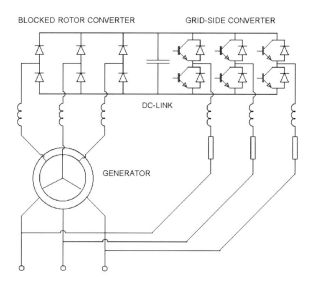

Figure 7.13   Illustration of the rotor converter blocking, with an opened rotor circuit.

This procedure allows fast demagnetisation of the rotor circuit (Akhmatov, 2002(a)). When the rotor current has decayed to zero, the rotor circuit is demagnetised. In this process, the magnetic energy of the current going through the rotor circuit inductance has transformed to the electric energy of the DC-link capacitor. As found in (Akhmatov, 2002(a)), the whole process from the moment when the IGBT-switches opens to when the rotor current has stopped takes some ms. In investigations of short-term voltage stability, this transient behaviour is less relevant and so the rotor current can simply be set to zero in the model when simulating the converter blocking (Akhmatov, 2002(b)).

When the rotor circuit has opened and demagnetised through the DC-link, but the stator terminals are still connected to the power grid, the generator cannot supply the active power to the grid, but the stator winding starts to absorb reactive power from the grid (Akhmatov, 2002(b)). When many generators operate with a grid-connected stator and an opened rotor, this may affect the grid voltage due to the quantity of the "reactors" absorbing the reactive power from the grid. To reduce the negative impact on the grid voltage, the wind turbines will disconnect shortly after the power frequency converters have blocked. Such reactive power absorption may only have a temporary character. When the power electronics converters have blocked, the sub-sequential disconnection of the wind turbines may have a positive impact on the power grid.

Figure 7.14 shows the simulated behaviour of the DFIG when the rotor converter blocks due to a grid voltage drop. In this case, the plotted curves are reached with application of the reduced converter model, e.g. with disregarded grid-side converter (as explained in **Section 7.4.3.3**). Although the applied converter model is simplified, the simulation result is correct and useful to illustrate the dynamic generator behaviour. In the simulated behaviour, the following events are noted, i.e. the moment when the:

1) Power grid is subject to the short-circuit fault.
2) Rotor converter has blocked and the rotor circuit has opened.
3) Generator trips.
4) Fault has cleared.

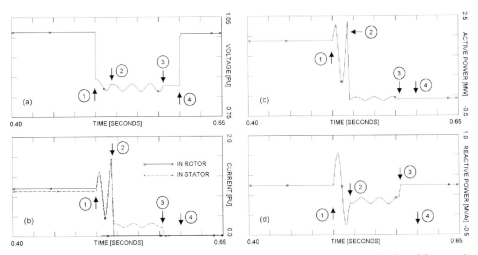

Figure 7.14 Simulated parameters of DFIG during a short circuit fault where the converter blocks and the rotor circuit opens: **(a)** – terminal voltage, **(b)** – currents in rotor and stator, **(c)** – active power and **(d)** – reactive power. Reprinted from Ref. (Akhmatov, 2002(b)), Copyright (2002), with permission from Multi-Science Publishing Company.

Note that when the voltage at the generator terminals has dropped to 0.8 p.u., the converter protective system is activated. The concept of variable-speed wind turbines equipped with DFIG and partial-load converters may be sensitive to such small voltage drops in the grid when fault-ride-through capability of the wind turbines is not required.

7.7.1.2 Rotor circuit "closes" through a crowbar

When the rotor converter is ordered to block, the IGBT-switches stop switching, the rotor converter trips and the rotor circuit is short-circuited, i.e. "closed", through a crowbar. When the rotor circuit is closed through a crowbar, the generator becomes a conventional induction generator. It continues to supply active power from the stator terminals to the grid and starts absorbing reactive power from the grid. The grid-side converter may be in operation awhile to balance the DC-link voltage. **Figure 7.15** shows the concept of this blocking sequence.

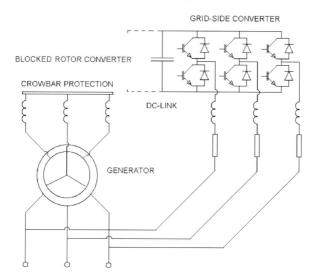

Figure 7.15   Illustration of the rotor converter blocking using a crowbar protection.

The initial generator rotor slip may be much larger than the initial slip of the corresponding conventional induction generator in the same operational point with respect to the active power. When the rotor circuit is closed through a crowbar, the generator becomes an over-sped induction generator which may cause significant reactive absorption, and large currents through the generator windings. Generator currents may contain excessive transients if the resistance of the crowbar is low. **Figure 7.16** illustrates the situation where the crowbar impedance is practically zero. When the crowbar resistance approaches 1000 times that of the rotor winding resistance, the current behaviour shown in **Figure 7.17** approaches the current behaviour reached for an opened rotor circuit (shown in **Figure 7.14**). This is not surprising because the opened rotor circuit may be interpreted as a rotor circuit with an infinitely large external resistance.

Simulation results are again reached using a reduced converter model, with a disregarded grid-side converter. In the plotted curves, the following events are noted, i.e. the moment when the:

1) Power grid is subject to the short-circuit fault.

2) Rotor converter has blocked and the rotor circuit has closed through the crowbar.

3) Generator trips.

4) Fault has cleared.

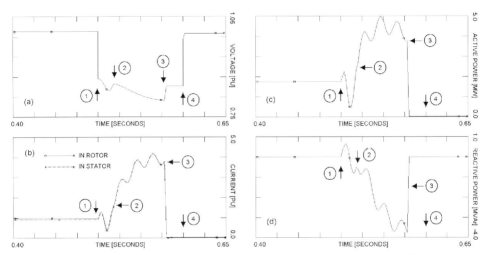

Figure 7.16 Simulated parameters of DFIG during a short-circuit fault where the rotor converter blocks and the rotor circuit has closed through a crowbar of infinitely small impedance: **(a)** – terminal voltage, **(b)** – current in rotor and stator, **(c)** – active power and **(d)** – reactive power. Reprinted from Ref. (Akhmatov, 2002(b)), Copyright (2002), with permission from Multi-Science Publishing Company.

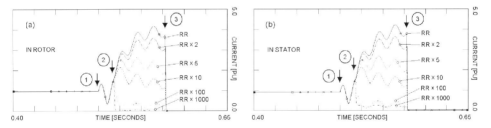

Figure 7.17 Simulated current in rotor and stator when DFIG is subject to a short-circuit fault, the rotor converter blocks and the rotor circuit is closed through a crowbar of finite resistance. RR stands for the rotor winding resistance, used for indication of the total rotor resistance including the crowbar. Reprinted from Ref. (Akhmatov, 2002(b)), Copyright (2002), with permission from Multi-Science Publishing Company.

Therefore, the crowbar must have a finite resistance value. **Figure 7.17** shows the simulated stator and rotor currents which are reached for various finite values of crowbar resistance. In the plots shown in **Figure 7.17**, the crowbar resistance is defined in relation to the resistance of the rotor winding itself. When the crowbar resistance increases, the stator and rotor currents are sufficiently reduced and the current transients are damped.

In the case of many generators maintaining grid-connection after their converters have blocked, the amount of the absorbed reactive power by the generators from the power grid may also be significant, and which can affect the grid voltage. To reduce such negative impact on the grid, wind turbines (without fault-ride-through) will disconnect shortly after their power electronics converters

7.8.1 Crowbar protection

The fault-ride-through solution with crowbar protection is based on the blocking and restarting sequences of the rotor converter and does not require significant increase in the power rating of the power electronics converter. The main idea behind this solution is that the wind turbines do not trip, but ride thorugh the grid disturbance, when the converters have blocked.

In (Pöller, 2003; Akhmatov, 2003(d)) this solution is described in detail for a case when the crowbar protection is applied at the same moment as the rotor converter blocks. When the rotor converter has blocked, the controllability of the rotor converter with regard to the independent control of the active and reactive power of the DFIG system is lost. When the rotor circuit of the generator is short-circuited through the crowbar, the generator operates as a conventional induction generator.

Just after the rotor converter has blocked, the generator may operate as a strongly over-sped induction generator and starts absorbing a significant amount of reactive power from the power grid. During this mode, pitch control is applied to prevent further excessive overspeeding of the induction generator (Akhmatov, 2002(b)). When the short-circuit fault is removed and the grid voltage and frequency are re-established by the control of conventional power plants, the rotor converter will be synchronised and restarted. After the rotor converter has restarted, the DFIG system returns to its regular operation with independent control of the active and reactive power as before the fault occurred. This fault-ride-through solution requires a successful restart of the rotor converter when the grid voltage recovers after the short-circuit fault. The main advantage of the fault-ride-through solution with the use of crowbar protection is that the wind turbines continue supplying active power to the grid. The main disadvantage is that the wind turbine generators absorb reactive power from the grid, which slows the process of the grid voltage re-establishment.

Note that blocking of a rotor converter does not introduce restrictions on uninterrupted operation of the grid-side converter. Therefore, the grid-side converter can be set to contribute to the reactive power control within the rated current range of the converter, when the rotor converter has blocked and the grid voltage is still low.

7.8.1.1 Strong power grids

In strong power grids, the grid voltage and frequency are re-established using the control of conventional power plants. The fault-ride-through operation of variable-speed wind turbines equipped with DFIG and partial-load frequency converters can be arranged employing crowbar protection. Loss of reactive power control of the DFIG systems during the rotor converter blocking sequence does not seem to be crucial to grid voltage re-establishment. In this situation, the rotor converter may wait to restart until the grid voltage is completely re-established and reactive power control provided by the grid-side converter is not required.

However some additional reactive power control of the grid-side converter may stabilise the voltage in the vicinity of the generator terminals and will be useful for the successful restarting of the rotor converter (Akhmatov, 2003(d)).

7.8.1.2   Weak power grids

In weak power grids, the reactive power control of the grid itself may be insufficient to support grid voltage re-establishment. In this case, a short-circuit fault may start voltage instability in the grid. If the grid voltage does not recover during a certain period, there can be a risk that rotor converters will not restart and the wind turbines disconnect from the grid. When the rotor converters have blocked at a voltage drop, the reactive power control provided by the grid-side converters will be relevant to support grid voltage re-establishment. However, the rotor converters must be able to restart relatively fast, perhaps before the grid voltage has completely re-established. When restarted, the rotor converter is applied to magnetise the wind turbine generators and control the reactive power in a more efficient way than the grid-side converter because the power rating of the converter-controlled generators is larger than that of the grid-side converters.

In weak power grids, the reactive power control of the power electronics converters can be combined with control provided by the dynamic reactive compensation units such as Static VAR Compensators, Statcoms, Synchronous Compensators and discrete components (capacitor banks). Dynamic reactive compensation units may be used for faster voltage re-establishment and faster restart of rotor converters used in DFIG based wind turbines.

7.8.1.3   Blocking and restarting sequences of rotor converters

The electric system of the DFIG system with the blocked rotor converter is shown in **Figure 7.19**. The protective system model of the DFIG and partial-load frequency converter is enabled in these simulations. The protective system monitors the grid voltage, the rotor current, the stator current, the DC-link voltage and other relevant parameters and orders the rotor converter to block when at least one of the monitored parameters exceeds the relay settings. When ordered to block, the rotor converter stops switching and trips. When the rotor converter blocks, the rotor circuit is short-circuited through the crowbar having the resistance R_{EXT} (Akhmatov, 2003(d)). The characteristic time of the crowbar activation is few ms (Akhmatov, 2002(b)).

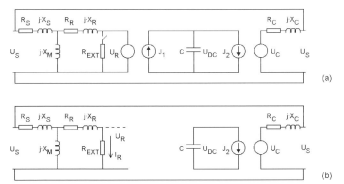

Figure 7.19   Schematic representation of DFIG and a converter system employing crowbar protection: **(a)** – cases of normal, uninterrupted operation of the frequency converters, synchronisation of rotor converter by $V_R = R_{EXT} I_R$, and restart of rotor converter, and **(b)** – operation with blocked rotor converter and operation of grid-side converter as a Statcom. Reprinted from Ref. (Akhmatov, 2003(d)), Copyright (2003), with permission from Multi-Science Publishing Company.

When the rotor converter has blocked, operation of the generator corresponds to the operation of a conventional induction generator with increased rotor resistance, $R_R = R_{R0} + R_{EXT}$. The resistance of the crowbar, R_{EXT}, has also a current-limiting function with regard to the operation of the rotor circuit. According to **Section 7.7.1.2**, the crowbar contributes to the dampening of excessive current transients in the rotor circuit, which may allow the faster restarting of the converter. During operations with the blocked rotor converter, the rotor converter control system continues monitoring the rotor current, the terminal voltage, the active power and the reactive power of the generator and the generator rotor speed. The rotor converter is waiting for the order from the control system to restart. The rotor converter de-blocking sequences are (i) synchronisation. and (ii) restart of the rotor converter.

When the grid voltage, frequency and, certainly, rotor current have returned to their acceptable ranges, the rotor converter starts synchronisation. At synchronisation, the rotor converter starts switching, whereas the crowbar having the resistance R_{EXT} is switched off.

Figure 7.20 presents the rotor converter control at synchronisation, which resets the PI-controllers by the actual (measured) values of the generator rotor speed, the active and reactive power, the (α,β)-components of the rotor current, etc. The (α,β)-components of the rotor voltage source are set to $u_{R\alpha} = R_{EXT} i_{R\alpha}$ and $u_{R\beta} = R_{EXT} i_{R\beta}$. This rotor converter operation sequence corresponds to inducing the rotor voltage source, U_R, following the voltage drop which would be present across the crowbar having resistance R_{EXT}. This operation is arranged using proportional controllers having gains R_{EXT} and (measured) rotor current, I_R, as the input of these controllers.

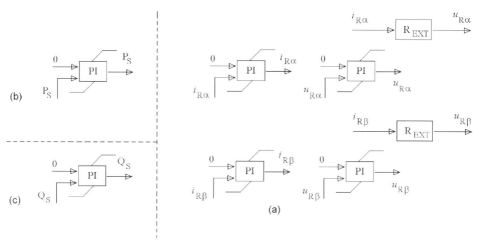

Figure 7.20   Generic control of a rotor converter during synchronisation: **(a)** – generic control of the rotor converter resetting the PI-controllers (cascade control), **(b)** – resetting of additional PI-controller producing the active power reference by speed, **(c)** – resetting of additional PI-controller producing the reactive power reference by voltage. Reprinted from Ref. (Akhmatov, 2003(d)), Copyright (2003), with permission from Multi-Science Publishing Company.

When the rotor converter waits for the right moment to restart, the control system monitors and resets the PI-controllers. When the synchronisation is complete, the control system of the rotor converter tries to go over to the regular operation mode shown in **Figure 7.7.a-b**. In other words, the rotor converter control attempts to restart. At the moment of the restart attempt, there can be a risk of that the rotor converter will again block. This rotor converter blocking can be caused by excessive transients in the rotor current due to insufficient synchronisation or insufficiently stable voltage in the vicinity of the wind turbine terminals. When the rotor converter attempts to restart at incomplete synchronisation, this may introduce sufficient error signals at the inputs of the PI- controllers of the rotor converter control. These error signals are the differences between the reference signals and the measured parameters. These signals will quickly be distributed throughout the proportional gains of the PI-controllers and affect operation of the rotor converter. The reference values must not be changed immediately at the moment of restart, but must be ramped from the initial reference values reached at synchronisation to the reference values set by the regular control.

When the rotor converter attempts to restart at insufficiently stable voltage, this may introduce significant voltage fluctuations in the vicinity of the wind turbine terminals. This may affect the rotor current in a way so that the rotor converter blocks again.

The risk of excessive transients in the rotor current at the moment of the rotor converter restart may also be reduced by independent restart of the control loops for the active power control producing the reference $u_{R\alpha}$ and the reactive power control producing the reference $u_{R\beta}$, e.g. when these two control loops are not restarted at the same time, but displaced in time. This displaces in time the rotor current transients occurring at the moments of restarts of these two respective control loops of the rotor converter and reduces the total value of the rotor current transients at the moment of the rotor converter restart.

7.8.1.4 Resistance of the crowbar

If the rotor circuit is simply closed through a low-resistance crowbar, the generator will operate as a strongly over-sped induction generator which is characterised by a large reactive current. In this case, excessive transients will be present in the rotor current (Akhmatov, 2002(b)). The crowbar, having the resistance R_{EXT}, must dampen such excessive transients in the rotor current when the rotor converter has blocked. The value of the crowbar resistance, R_{EXT}, is a design parameter and influences also the reactive power absorption of the induction generator from the power grid. Thus, the following considerations are made when choosing the value of R_{EXT}.

1) Transients in the rotor current must efficiently be damped before the converter synchronisation and the converter restart will take place.
2) The crowbar resistance must not be much large to minimise excessive transients in the rotor current when the crowbar will trip and the rotor converter restarts.
3) The crowbar resistance must be set to minimise the reactive power absorption of the generator from the power grid.
4) The value of R_{EXT} must be chosen in relation to the other electrical parameters of the generator.

In Ref. (Akhmatov, 2003(d)) it is suggested applying the crowbar resistance $R_{EXT} = 20 \cdot R_{R0}$, where R_{R0} is the resistance of the generator rotor winding itself.

7.8.1.5 Operation of the grid-side converter

When the rotor circuit is short-circuited through the crowbar, the rotor converter is tripped and the active power cannot be transferred between the rotor circuit and the power network via the DC-link of the converter system. The grid-side converter is kept in operation and set to control the DC-link voltage.

From the moment of synchronisation, the rotor converter is taken in operation and the active power $P_R = V_R conj(I_R)$ is exchanged between the rotor circuit and the grid-side converter through the DC-link. Then, the active power will be supplied to, or absorbed from, the power grid at the grid-side converter terminals from the moment of synchronisation to balance the DC-link voltage.

When the regular control is not sufficient to keep the DC- link voltage within a desired range, additional arrangements (such as choppers in the DC- link activated at excessive over-voltage) can be applied. This is discussed in **Section 7.8.2.1**.

Furthermore, the grid-side converter can be set to control the reactive power during severe grid events. This requires the coordination of control between the grid-side and the rotor converter.

7.8.1.6 Reactive power control coordination

When the rotor converter has blocked (the rotor circuit is closed through the crowbar) or when the rotor converter is synchronised, the wind turbine generator absorbs the reactive power from the power grid. This is necessary to excite the generator.

Since the reactive power control of the rotor converters is inactive, the reactive power demands of the generators may be covered by the use of a dynamic reactive compensation unit connected at the common connection point of the windfarm. The dynamic reactive compensation unit can be commissioned as (i) discrete components such as a capacitor bank, (ii) components with continuous control of the reactive power such as SVC, Statcoms and Synchronous Compensators and (iii) as a combination of discrete and continuously controlled components. When applied together with a large windfarm, the superior control system of the windfarm (a so-called windfarm controller) can be set to access the reactive power control of the dynamic reactive compensation unit and order the unit to support the grid voltage at short-circuit faults in the power grid.

Application of Statcoms and Synchronous Compensators may reduce voltage drop at the terminals of the large windfarm during a grid fault. This may therefore reduce the magnitude of the rotor current transients and prevent the rotor converter blocking.

The partial-load frequency converters of the wind turbines can also be set to control the reactive power and to support the grid voltage. This may reduce demands on the incorporation of dynamic reactive compensation units in the power grid (Akhmatov, 2002(b)). Application of the reactive power control of the partial-load frequency converters require (i) a fast restart of the rotor converters and (ii) the use of grid-side converters to control the reactive power in the situations when the rotor converters have blocked (Akhmatov, 2003(d)).

Since the power rating of a grid-side converter is smaller than that of a converter controlled generator, the value of the reactive power control provided by the grid-side converter is also smaller. Therefore the reactive power control must come from the control of the rotor converter, whereas the reactive power control of the grid-side converter is used in situations where the rotor converter has blocked.

During short-circuit faults in the grid, the grid-side converter must maintain uninterrupted operation. Operation of the grid-side converter becomes then analogous to a Statcom, see **Figure 7.19.b**, which is why it is feasible to use this converter to control the reactive power. This control can be arranged with the control systems shown in **Figure 7.7.d-e**. When the rotor converter has blocked, the grid-side converter can be set to control the reactive power for faster re-establishment and stabilisation of the grid voltage. This will prepare the fast restart of the rotor converter.

To avoid competition between the reactive power control provided by the (restarted) rotor converter and the grid-side converter, the rotor converter controls the reactive power in any situation except when this converter has blocked or is synchronising. The grid-side converter is set to control the reactive power only when the rotor converter has blocked or is synchronising. In all other situations, the grid-side converter is kept reactive neutral, i.e. it does not exchange reactive power with the grid (Akhmatov, 2003(d)).

Keep in mind that the power rating of the grid-side converter may be around 25% that of the generator rated power. When the rated power of the windfarm is 160 MW, the grid-side converter can, presumably, control the reactive power in the range of 40 MVAr. This corresponds to the incorporation of a Statcom with the power rating of 40 MVA. This is not much compared to the rated power of the windfarm, but this "Statcom" feeds directly into the local (offshore) power network of the windfarm. As the "Statcom" is placed at the wind turbine terminals, the unit can be very efficient to stabilise the voltage in the vicinity of the wind turbine terminals and contributes to the fast and safe restarting of the rotor converters.

7.8.1.7 Simulation example

Use of crowbar protection for fault-ride-through is modelled for a large offshore windfarm consisting of eighty variable-speed wind turbines. The windfarm model with representation of the internal cable network, the high-voltage cable and the connection point on land is described in **Section 5.7.1**. **Figure 7.21** shows the simulated behaviour of wind turbine WT 01 with the use of crowbar protection of the rotor converter. From the start of the simulation ($t = 0$) to time $t - T_1$, the power grid is in normal operation.

At time $t = T_1$, a short-circuit fault occurs in the transmission power grid and the grid voltage drops to about 0.25 p.u. The voltage drop initiates current transients in the stator and rotor circuits of the DFIG, as well as in the grid-side converter mains, and DC-link voltage fluctuations (Akhmatov, 2002(b); Akhmatov, 2003(d)). Excessive current transients and DC-link voltage fluctuations cause the rotor converter to block.

At time $t - T_2$, the rotor converter blocks due to excessive transients in the rotor current, which is immediately followed by short-circuiting of the rotor circuit through the crowbar with the resistance R_{EXT}. When the rotor converter has blocked, wind turbine WT 01 operates as a pitch-controlled wind turbine equipped with a conventional induction generator with enlarged rotor resistance; this prevents the wind turbine from excessive over-speeding.

The grid-side converter is set to operate as a Statcom, and is set to balance the DC-link voltage, to control the reactive power and to support the grid voltage near to the wind turbine terminals. The

bar protection (the blocking and restarting sequences of the rotor converters) is robust enough to be used in large offshore windfarms.

As expressed in Ref. (Slootweg, 2003), well-tuned power electronics converters will not interact with each other, independently of the operational situation in the power grid.

7.8.2 Current control and limitation

Excessive transients in the rotor current are the main reason of the rotor converter blocking. When uninterrupted operation of the rotor converter during a voltage drop is required, early converter restart and rotor current control and limitation can be applied (Akhmatov, 2002(b)). When the wind turbine is subject to a short-circuit fault, the magnitude of the phase currents in the rotor windings increases rapidly. The rotor converter control registers such abnormal operation and orders the rotor converter to block.

Fast elimination of excessive transients in the rotor current can be gained by opening the rotor circuit and de-excitation through the diode bridge of the rotor converter, or by applying a sufficiently large crowbar switched to the rotor circuit, see **Section 7.7.1**. When excessive transients are eliminated, the IGBT-switches start switching again and the rotor converter is restarted. However, the switching pattern is changed to limit the current in each single phase of the rotor circuit (Akhmatov, 2002(b)). The duration of the converter blocking sequence is short compared to the period of the short-circuit fault and can, under certain assumptions, be disregarded in simulations. **Figure 7.23** presents the computed results for this control feature when converter blocking is disregarded.

When the rotor converter has restarted, the active and the reactive power references of the rotor converter control are guided to meet the requirements for a given power system.

When voltage support is required, the active power reference, $P_{E,REF}$, is reduced and the reactive power reference, $Q_{S,REF}$, increased during the voltage drop and post-fault voltage re-establishment. The reactive power reference, $Q_{S,REF}$, can be controlled by the grid voltage as shown in **Figure 7.8.c** whereas the active power reference is kept below the upper limit value of

$$P_{E,REF(MAX)} = \sqrt{S_{RAT}^2 - Q_{E,REF}^2} \ .$$

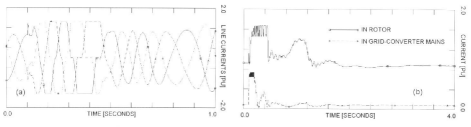

Figure 7.23   Rotor current limitation applied to prevent rotor converter blocking: **(a)** - single phase currents in rotor during a grid voltage drop, **(b)** - current in the rotor circuit and grid-side converter mains during a longer period. Presented in Ref. (Akhmatov, 2002(b)), Copyright (2002), with permission from the copyright holder.

When frequency support is required, the active power reference, $P_{E.REF}$, is maximised or controlled by the grid frequency, whereas the reactive power reference is controlled by the terminal voltage within the converter rating range as $Q_{S,REF(MIN,MAX)} = \mp\sqrt{S_{RAT}^2 - P_{E.REF}^2}$.

When voltage and frequency supports are both required, it is necessary to share the converter controllability between the active power and the reactive power controls according to an algorithm corresponding to the requirements of a given power system.

For efficient control of the active and reactive power, the voltage at the DFIG terminals should not drop below approximately 0.4 p.u. (Akhmatov, 2002(b)). If the residual voltage at the DFIG terminals becomes smaller than 0.4 p.u., there can be a risk that the control provided by the rotor converter will not be sufficient enough, although the rotor converter has restarted and works in the current limiting mode. The controllability of the rotor converter is limited by the rotor current and voltage limits. Therefore, this fault ride-through feature may require that the power rating of the rotor converter is increased. The limits of the rotor current and the induced rotor voltage, $i_{R\alpha.MIN}$, $i_{R\beta.MAX}$, $u_{R\alpha.MIN}$ and $u_{R\beta.MAX}$, of the respective PI-controllers of the rotor converter cascade control in **Figure 7.8.a** must then be increased.

When frequency support is required, the limits of the active rotor current are set to the rated current of the rotor converter, $i_{R\alpha(MIN,MAX)} = \mp i_{RAT}$, whereas the limits of the reactive current are dynamically adjusted according to $i_{R\beta(MIN,MAX)} = \mp\sqrt{i_{RAT}^2 - i_{R\alpha}^2}$. When voltage support is required, the limits of the reactive rotor current are set to the rated current of the rotor converter, $i_{R\beta(MIN,MAX)} = \mp i_{RAT}$, whereas the limits of the active current are dynamically adjusted according to $i_{R\alpha(MIN,MAX)} = \mp\sqrt{i_{RAT}^2 - i_{R\beta}^2}$. When frequency and voltage support are required at the same time, the active and reactive current limits can be guided by a certain algorithm to reach the required control properties.

The limiting conditions applied to the dynamically controlled current and voltage in the rotor circuit are:

$$\begin{cases} i_{R\alpha.MIN} \leq i_{R\alpha} \leq i_{R\alpha.MAX}, \\ i_{R\beta.MIN} \leq i_{R\beta} \leq i_{R\beta.MAX}, \end{cases} \qquad (7.29)$$
$$\begin{cases} u_{R\alpha.MIN} \leq u_{R\alpha} \leq u_{R\alpha.MAX}, \\ u_{R\beta.MIN} \leq u_{R\beta} \leq u_{R\beta.MAX}, \end{cases}$$

These limiting conditions must be fulfilled during rotor converter (low-voltage) operation. Note that the current limitation may introduce distortion, i.e. higher harmonics of the fundamental grid frequency, in the rotor and grid-side converter current. However this is not relevant for investigations of short-term voltage stability.

Although the increase of power rating of the rotor converter contradicts with the idea of the use of smaller rotor converters, this arrangement reduces the risk of generator operation with the rotor converter blocking and improves the fault-ride-through capability of wind turbines. Note that when rotor converter control is active, the converter is able to control the reactive power within the converter rating range and support the grid voltage in the vicinity of the DFIG terminals. This reactive power control provided by the rotor converter may reduce the voltage drop at the DFIG terminals as well.

The fault-ride-through solution using rotor current control and limitation (Akhmatov, 2002(b)) is an example of how to maintain rotor converters of such variable-speed wind turbines in operation and support the reactive power and grid voltage during and after short-circuit faults in the grid. When applying this control solution, two important tasks must be investigated.

1) Interaction between the large windfarm and power grid, including the short-term voltage stability of the grid and the fault-ride-through capability of the wind turbines.
2) Risk of mutual interaction between the wind turbines within the windfarm and the interaction between the control systems of the partial-load frequency converters.

For "pedagogical" reasons, the rotor converter using current control and limitation can additionally be equipped with crowbar protection. Crowbar protection can be activated if the current control and limitation feature is not sufficient for maintaining the unblocked operation of the rotor converter at the voltage drop, or if such a control feature fails (for example, during extreme long and efficient short-circuit faults).

Ref. (Høgdahl and Nielsen, 2005) presents an example on the DFIG- based wind turbine from the manufacturer Vestas Wind Systems with the rotor converter operating at significant voltage drops. In the test, the voltage drop was down to 0.5 p.u. during 500 ms. The active power of the 2 MW wind turbine being at rated operation prior to the voltage drop showed a drop to 700 kW during the voltage drop. The reactive power increased from 0 kVAr prior to the voltage drop to 300 kVAr during the voltage drop, which is a part of the low-voltage fault-ride-through control. In this concept, the rotor converter control is changed during the grid disturbance and set to deliver the reactive power to the power network. Ref. (Høgdahl and Nielsen, 2005) does not provide detailed information on the converter control, but confirms that the DFIG based wind turbines are capable of supporting the grid voltage during, and after, a short-circuit fault.

7.8.2.1 Grid-side converter control

Grid-side converters must keep the DC-link voltage within the required range of operation and, presumably, provide additional reactive power control in emergency situations.

During a grid voltage drop and the blocking of the rotor converter, the DC-link voltage may fluctuate. The DC-link voltage however may not increase more than 10% of the rated value (Petesson, 2005). During the whole operation with the rotor converter blocked and after the converter restart, the grid-side converter controls the DC-link voltage employing a regular control loop shown in **Figure 7.8.d**. Additionally, a breaking chopper can be installed at the DC-link (Petersson, 2005). This breaking chopper acts as a load dump at excessive over-voltage in the DC-link, which cannot be removed by the regular converter control. This improves the ride-through capability of the converter system at a significant grid voltage drop, but is limited by the heat generated by the braking resistor and not applicable for long durations.

When the rotor converter controls the reactive power, the reactive power reference of the grid-side converter is usually kept at zero. This arrangement is to avoid possible conflict between the converter controls, if both converters wished to control the grid voltage nearby the DFIG terminals. When the reactive power control provided by the rotor converter is limited by for example the converter current rating, the grid-side converter can be useful to contribute to the grid voltage support. Then, the converter control coordination is required. For this control coordination, the main slow voltage control can be arranged by the rotor converter. The additional fast voltage control is arranged by the grid-side converter. This control principle is illustrated in **Figure 7.24**. The grid-side converter supplies the reactive power, but still within its rating, when the rotor side converter operation is restricted.

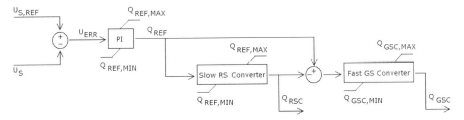

Figure 7.24 Reactive power control coordination with uninterrupted converter operation. Here Q_{RSC} and Q_{GSC} denote the reactive power supplied by the stator mains of the DFIG and by the grid-side converter, respectively.

Note that the terms slow and fast are very relative in this context. The characteristic response times of both converters are still within tens to hundreds of milliseconds.

7.8.2.2 Simulation example

In this simulation example, a large offshore windfarm is commissioned at Omø and connected to the transmission grid of Eastern Denmark at the rated grid voltage of 132 kV. The large offshore windfarm contains pitch-controlled, variable-speed wind turbines equipped with DFIG and partial-load frequency converters. The rated power of the windfarm is 150 MW (Akhmatov, 2002(b)).

In these simulations, a realistic model of the eastern Danish power system containing detailed models of conventional power plants with control, models of consumption centres and models of dispersed generation units such as combined heat-power (CHP) units and local wind turbine clusters is applied. **Figure 7.25** presents a sketch of the eastern Danish power grid. Local wind turbine clusters are represented as re-scaled equivalents of stall-controlled, fixed-speed wind turbines equipped with conventional induction generators. CHP units are represented by synchronous generators without any specific control. The models of the wind turbines and the CHP units contain also models of protective relays, which monitor the grid voltage, machine current, grid frequency and speed and automatically disconnect the wind turbines and the CHP units when the relay settings are exceeded.

The large offshore windfarm at Rødsand 1 / Nysted commissioned in 2003 and the considered windfarm at Omø are represented by their re-scaled, single-machine equivalents on the assumption that all the wind turbines in the windfarms are in rated operation (Akhmatov and Knudsen, 2002). The rated operation gives the worst case with regard to maintaining of short-term voltage stability in the case of fixed-speed wind turbines at Rødsand 1 / Nysted. The rated operation will also give the worst case with regard to the fault-ride-through capability of variable-speed, DFIG-based wind turbines at Omø. Use of such a detailed model of the Eastern Danish transmission grid with representation of the control of conventional power plants and the impact from dynamic consumption centres and the dispersed generation units produces a more realistic approach with regard to the operation of the power electronics converters of the wind turbines at Omø.

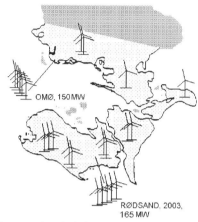

Figure 7.25 An example of the incorporation of wind power in Eastern Denmark including a windfarm at Omø. Reprinted from Ref. (Akhmatov, 2002(b)), Copyright (2002), with permission from Multi-Science Publishing Company.

The disturbance is a short-circuit fault of 100 ms of duration at a selected node in the transmission grid that is not far from the on-land connection point of the offshore windfarm at Omø. When the short-circuit fault is cleared, the transmission lines connected to the faulted node are tripped making the power system weaker.

The power electronics converters of the wind turbines in the Omø offshore windfarm are with a fault-ride-through solution applying the current control and limiting feature (Akhmatov, 2002(b)). **Figure 7.26** presents simulation results for the variable-speed wind turbines at Omø. When the grid voltage drops, the active power reference, $P_{E,REF}$, is reduced whereas the reactive power reference, $Q_{E,REF}$, is controlled by the grid voltage, see **Figure 7.8.c**. The Omø offshore windfarm controls the reactive power and supports grid voltage re-establishment as known from operation of Synchronous Compensators. Establishment of a separate unit of dynamic reactive compensation is not necessary in this case.

Note that the Omø offshore windfarm supplies reactive power and contributes to voltage re-establishment during the voltage drop. However, several local wind turbines have tripped during the fault producing a surplus of reactive power in the grid (Akhmatov, 2002(b)). When the grid voltage has re-established, the Omø offshore windfarm absorbs this reactive power surplus and keeps the grid voltage within a desired range.

7.8.2.3   Mutual interaction

The mutual interaction issue relates to, if the fast-acting control of the power electronics converters of many variable-speed wind turbines within a large (offshore) windfarm may interact with each other. The investigation of Ref. (Akhmatov, 2002(b)) focused on the performance of a large off-shore windfarm applying the rotor converter control with the current control and limitation (the fault-ride-through solution). The large offshore windfarm is divided into eight sections with a grid-connected single-machine equivalent in each section. The rated power of a single-machine equivalent is 20 MW, which represents a number of variable-speed wind turbines connected to the given section of the windfarm.

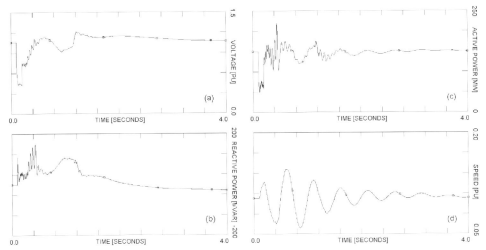

Figure 7.26   Simulated dynamic behaviour of the Omo offshore windfarm when subject to a short circuit fault in the transmission grid: **(a)** – terminal voltage, **(b)** – reactive power, **(c)** – active power and **(d)** – generator rotor speed. Reprinted from Ref. (Akhmatov, 2002(b)), Copyright (2002), with permission from Multi-Science Publishing Company.

Figure 7.27 illustrates the model set-up of the large offshore windfarm of Ref. (Akhmatov, 2002(b)). This model recognises that the sections of the large windfarm may shadow each other from incoming wind and therefore have slightly different initial operation points. As assumed, the active power reduces by 0.5 MW per a section.

The short-circuit fault has a duration of 100 ms and simulation results are shown in **Figure 7.28**. As can be seen, the wind turbine sections show a coherent response to the short-circuit fault. The wind turbine sections do not oscillate against each other or against other control systems in the power grid. Since the variable-speed wind turbines are with well-tuned control systems using power electronics converters, there is no mutual interaction between the converter control systems during such grid disturbances.

7.8.3   Dampening of shaft torsion oscillations

As reported in Ref. (Akhmatov, 2002(b)), undamped torsion oscillations may be excited in shaft systems of wind turbines equipped with DFIG and partial-load frequency converters when only the cascade-control loops of the rotor converter (shown in **Figure 7.8.a**) are applied without any damp-ening arrangement. **Figure 7.29** showns an example on such undamped torsion oscillations gained for a realistic representation of the rotor converter control (as in **Figure 7.8.a**).

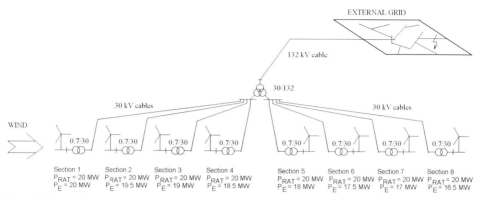

Figure 7.27   Model set-up of the large offshore windfarm with eight sections. Reprinted from Ref. (Akhmatov, 2002(b)), Copyright (2002), with permission from Multi-Science Publishing Company.

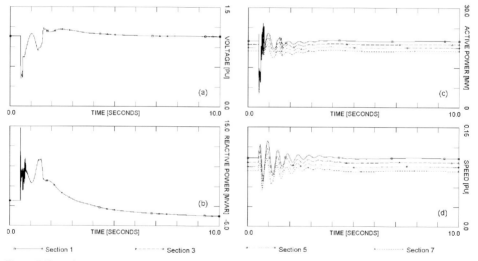

Figure 7.28   Coherent response of a large offshore windfarm modelled as eight individual wind turbine sections: **(a)** - terminal voltage, **(b)** - reactive power, **(c)** - active power and **(d)** - generator rotor speed. Reprinted from Ref. (Akhmatov, 2002(b)), Copyright (2002), with permission from Multi-Science Publishing Company.

To dampen such undesired torsion oscillations in the shaft system, a dampening control loop is added to the rotor converter control, see **Figure 7.8.c**. In this additional control loop, the active power reference, $P_{E,REF}$, is set to oscillate against the measured generator rotor speed, ω_G, which is found efficient to dampen the shaft torsion oscillations (Akhmatov, 2002(b)). The result is seen in **Figure 7.26.d** as a well-damped torsion mode in the shaft system.

Functionality of this additional control loop is similar to application of Power System Stabilizers (PSS) to conventional power plants equipped with synchronous generators. In the case of conventional power plants, the PSS is set to dampen power oscillations between different machines or between different areas of a large power grid. It is thinkable that the rotor converter control of DFIG may also be applied for the same purpose. In this case, it may have active power as an input signal sent through a number of wash-out filters before it reaches the additional dampening control in the rotor converter control system.

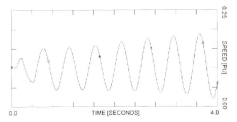

Figure 7.29 Generator rotor speed with insufficient dampening of shaft torsion oscillations (without PI-controller producing the $P_{E.REF}$ signal shown in **Figure 7.8.c**) resulting in self-excitation of the shaft system of the DFIG. Reprinted from Ref. (Akhmatov, 2002(b)), Copyright (2002), with permission from Multi-Science Publishing Company.

The parameters of the PI-controller in **Figure 7.8.c** applied for the dampening of shaft torsion oscillations can be chosen knowing the natural frequency of the shaft system and in agreement with other parameters of the converter control (Akhmatov, 2002(b)). Note that the predicted efficiency of the rotor converter control to dampen torsion oscillations in the shaft system may depend on the model complexity of the converter system. As reported in Ref. (Akhmatov, 2003(e)), simplified models of converter systems neglecting the dynamics of the grid-side converter with the DC-link may predict inadequate results for the torsion oscillation dampening.

Let the gain of the PI-controller in **Figure 7.8.c** producing the $P_{E.REF}$ signal be denoted as K_{PREF}. Simulated curves of generator rotor speed at different gains K_{PREF} are plotted in **Figure 7.30**. The computations are made when the converter system is modelled with complete representation of the rotor and grid-side converter control with the DC-link. For reaching efficient dampening of the torsion oscillations in the shaft systems, the gain K_{PREF} must be approximately 5 p.u. which corresponds to the value $K_{PREF} = K_4$ in this particular case.

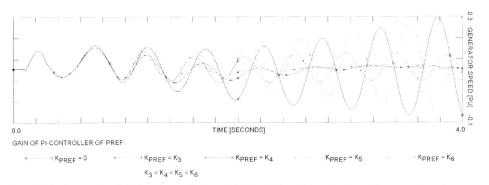

Figure 7.30 Dampening of shaft torsion oscillations at different values of the gain K_{PREF}. The converter system represents the rotor converter control, grid-side converter control and the DC-link. Reprinted from Ref. (Akhmatov, 2003(e)), Copyright (2003), with permission from Multi-Science Publishing Company.

Note that when the reduced converter model disregarding the grid-side converter and DC-link is applied, se **Figure 7.9**, inadequacy of this reduced converter representation will be exposed by the following.

1) The reduced model predicts that the gain K_{PREF} must be chosen in the range of K_6 to reach the best dampening performance of the converter control.
2) In contrast, the gain K_{PREF} in the range of K_4 will predict undamped shaft oscillations.

As the physical converter systems always contain the rotor and grid-side converter with a DC-link, reduced models of the converter systems may introduce inaccuracies with regard to the dynamic performance of the converter control.

Reduced models of the converter systems disregarding the dynamics of the grid-side converter control with the DC-link cannot be applied for tuning the converter control parameters to dampen shaft torsion oscillations. However, the reduced models of the converter control may be applied in investigations of short-term voltage stability, but at percussions with regard to the predicted results.

7.8.4 Solutions of wind turbine manufacturers

All wind turbine manufacturers offer specific solutions for maintaining fault-ride-through operation of their variable-speed wind turbines equipped with DFIG and partial-load frequency converters. Solutions must comply with the Grid Codes of the national system operators and are among the most significant competitive parameters playing a part in awarding of large-scale projects (Bolik, 2003). All the solutions and their specific details are either patented or kept classified as the competition between manufacturers is tough. Several wind turbine manufacturers apply a fault ride-through solution using crowbar protection described in **Section 7.8.1**. The manufacturer Vestas Wind Systems have announced the Advanced Grid Operation (AGO) concept which is a series of solutions based on different control regimes used by power electronics converters to ride through short-circuit faults without using crowbar protection (Eek et al., 2004; Høgdahl and Nielsen, 2005).

The wind turbine manufacturer GE Wind Energy has developed fault-ride-through solutions termed the Low-Voltage Ride-Through (LVRT[TM,11]). The LVRT[TM] solutions address the voltage stability concerns enabling wind turbines to ride through the short-circuit faults without disconnection and stopping (Miller, 2003). The wind turbine generators feed the reactive power into the power grid during such short-circuit faults and contribute to the grid voltage re-establishment. The control system of the power electronics converters is designed to deliver ride-through capability at or above 15% grid voltage. The LVRT[TM] also remains engaged until after the short-circuit fault is cleared, providing support to bring the system back to normal operation. The LVRT[TM] optimises all of the electronically controllable components in the wind turbines such as the power electronics converters, the pitch control, the motors and the pumps so the wind turbine generators maintain high availability at low voltages. The response of the variable-speed DFIG-based wind turbines equipped the LVRT[TM] to the single phase-to-ground as well as to the 3-phase short-circuit fault is presented in **Figure 7.31** and in **Figure 7.32**, respectively, with permission from the manufacturer

[11] LVRT[TM] is a trade mark of the wind turbine manufacturer GE Wind Energy.

GE Wind Energy. Presently, the LVRT[TM] is applied in the GE wind turbines commissioned in several windfarms world-wide and specifically in the Arklow offshore windfarm in Ireland illustrated in the photographs shown in **Figure 1.3** and **Figure 2.2**.

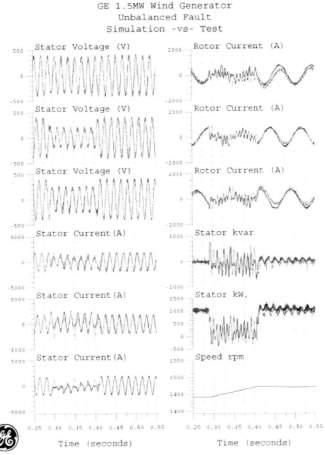

Figure 7.31 The LVRT[TM] simulation and test of the 1.5 MW variable-speed wind turbine equipped with a DFIG and a partial-load frequency converter from the manufacturer GE Wind Energy with regard to a single phase-to-ground short-circuit fault, Copyright (2003), with permission from GE Wind Energy.

The ability of variable-speed wind turbines to provide voltage regulation and fault-ride-through capability is in constant improvement. The target is to operate the large (offshore) windfarms as the wind power plants having similar controllability as for conventional power plants.

7.9 Summary

Variable-speed wind turbines equipped with DFIG and partial-load frequency converters are a relevant concept for large (offshore) windfarms. The use of variable-speed operation and pitch control allows for better optimisation of power output according to incoming wind. The use of power

electronics converters allows independent control of active and reactive power similar to the operation of synchronous generators when the power grid is at normal operation. In the case of variable-
speed DFIG-based wind turbines, the generators are excited through the rotor circuits applying the
control of the rotor converters, but not from the power grid through the stator terminals as in the
case of conventional induction generators. Such variable-speed wind turbines may be set to control
the reactive power within a given range and support the grid voltage.

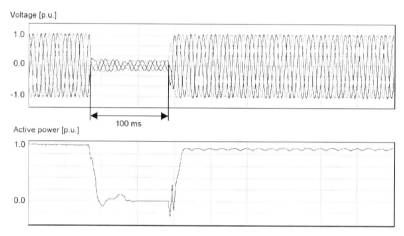

Figure 7.32   The LVRT™ simulation and testing of a variable-speed wind turbines equipped with DFIG and partial-
load frequency converters from the manufacturer GE Wind Energy with regard to a 3-phase short-circuit
fault, Copyright (2003), with permission from GE Wind Energy.

DFIG systems contain generators controlled by partial-load frequency converters. Application of
partial-load frequency converters of smaller power ratings is the main advantage of this wind turbine concept as smaller converters reduce the size and the cost of converters. However, such power
electronics converters are sensitive to excessive transients in the rotor current and thermal overloads and may block when the power grid is subject to a short-circuit fault.

When the rotor converter has blocked, the IGBT switches of the rotor converter stop switching
and the converter trips. The rotor converter may block by excessive transients in the rotor current,
excessive over-voltage in the DC-link, under- and over-voltage in the power grid, fluctuations of the
grid frequency, etc. The converter protective system monitors the relevant parameters of the generator and the converter itself and orders the rotor converter to block when at least one of the monitored parameters exceeds its relay settings. When blocked, the controllability of the rotor converter
is lost. Blocking of the converters may also lead to the disconnection and stopping of the wind turbines when fault ride-through solutions are not applied.

The bottom line is that the converter action, control, protection and ability to handle short-circuit
faults will decide the operation of the wind turbine and its generator during such short-circuit faults
in the power grid.

Application of converter control implies that wind turbines may operate within a relatively large
range of generator rotor speed and excessive over-speeding does not relate to voltage instability in
this wind turbine concept. However, application of power electronics converters can introduce other

challenges in this concept such as a restriction on converter control during the fault or during grid operations at low voltages. Therefore, the fault-ride-through capability of variable-speed wind turbines with DFIG is a key issue.

The fault-ride-through operation can be gained in several ways. The generator may, for example, maintain uninterrupted operation and ride through the grid faults whereas the rotor converter has blocked. In this situation, the rotor circuit of the generator may be short-circuited through a finite-resistance crowbar at the same moment as the rotor converter has blocked. The generator continues operating as a conventional induction generator, supplying active power to, and absorbing reactive power from, the grid. When the short-circuit fault is well-cleared and the grid voltage and frequency have re-established, the rotor converter restarts and the crowbar is removed. This solution is called "crowbar protection", applied by some wind turbine manufacturers.

Note that the grid-side converter may be set to supply reactive power in the range of the converter current rating and so support grid voltage re-establishment. This additional control of reactive power is useful in situations when the rotor converter has blocked due to a grid fault. The current rating of the grid-side converter is however smaller than that of the complete DFIG- system.

The main disadvantage of crowbar protection is that the generator must be magnetised from the power grid when the rotor converter has blocked. Solutions that do not require blocking of the rotor converter during a relatively long period are preferred. When the rotor converter restarts at low-voltage operation of the grid and rides through the fault, the DFIG system may be set to supply reactive power and support grid voltage re-establishment when the fault is cleared. Such solutions may require the use of the rotor converters with enlarged power ratings.

Solutions using a rotor converter to ride through grid faults are already available on market. For example, the wind turbine manufacturer GE Wind Energy offer the exceptional LVRT[TM] control system allowing their wind turbines to ride through grid faults and supply reactive power to the grid.

Wind turbine models applied in investigations of short-term voltage stability must predict the accurate response of converters to short-circuit faults. As the rotor converters are sensitive to, and may block due to, current transients, the generator model must be a transient, fifth-order model of induction generators, e.g. with representation of fundamental-frequency transients in the rotor current. In investigations of power system stability of large power systems with a significant number of grid components, a reduced, third-order model of induction generators is commonly applied (often as a standardised model of the available simulation tool). When the more correct, transient fifth-order model is not available, the third-order model can be applied under guidance of the wind turbine manufacturer. Obviously, the third-order model cannot predict rotor converter blocking due to excessive current transients. Instead, rotor converter blocking can be initiated in the model by a voltage drop below a certain value (and during a certain period), given by the wind turbine manufacturer. Typically, the voltage drop can be to about 0.8 p.u. and the drop period equivalent to a grid-frequency period (20 ms in Europe at 50 Hz).

In some simulation tools, the representation of the grid-side converter with a DC-link is neglected and rotor active power is "directly" fed into the power grid model. This assumption can be justified for investigations of power system stability, but a careful interpretation of the results is required. The reduced converter model, e.g. with only the representation of rotor converter control, may produce inadequate dynamic performance of the converter, for example, inadequate dampening characteristics of the converter system. Furthermore, the DC-link voltage is among the parameters monitored by the converter protective system.

The reduced models of the generator and the converter control may be applied in power stability investigations, but at percussions with regard to the predicted results.

Commissioning of a wind turbine. Photo copyright GE Wind Energy. Reproduced with permission from GE Wind Energy.

8 Induction generators with full-rating converters

As the rated power of modern wind turbines shows a tendency to increase, rotor blade lengths tend to increase whereas rotor speeds decreases, see **Section 3.1.2** and **Section 3.1.3.2**. For example, a wind turbine with 5 MW rated power may have a blade length of more than 60 m and a rated rotor speed in the range of 10 rev./min. When the generator is with two pole-pairs, the mechanical gear ratio must be (1500 rev./min. /10 rev./min.) =150 to provide a direct AC connection of a 5 MW wind turbine generator to the power grid with a fixed frequency of 50 Hz. Note that the mechanical gear ratio of modern 2 MW wind turbines is up to 100, see **Section 3.2.2.1**.

Such significant increases in the mechanical gear ratio introduce a challenge for the design and construction of gearboxes applied in MW- class wind turbines. Size, weight and friction losses also increase when the rating and gearbox ratios increase. Alternatively, full-rating frequency converters for the AC/DC/AC connection of the wind turbine generators to the power grid can be used.

The concept of converter-connected induction generators employing full-rating frequency converter has been used by the manufacturer Siemens in their MW- class wind turbines (2.3 MW and 3.6 MW announced in 2005). A scheme of a converter-connected induction generator is shown in **Figure 8.1.a**. In this wind turbine concept, a large mechanical gearbox can be replaced by a smaller mechanical gearbox and a full-rating frequency converter. The speed conversion from the wind turbine rotor speed to the rated grid frequency is shared between the gearbox and the frequency converter. This reduces the total power losses caused by the mechanical speed conversion. As the no-load power losses are reduced, wind turbines may start operations at lower wind speeds.

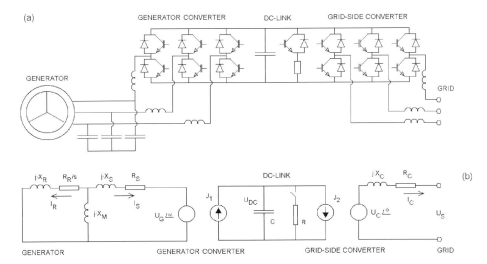

Figure 8.1 A converter connected induction generator: **(a)** - scheme with VSC converters, **(b)** - simplified representation in steady-state computations.

The application of full-rating frequency converters decouples the rotor speed from the grid frequency which is kept fixed to the rated value and, then, allows variable-speed operation of the wind turbine. Wind turbines with such converter-connected induction generators are also pitch controlled.

Optimisation of the power output by the variable-speed operation is described in **Section 3.1.3** whereas pitch control is present in **Section 3.1.4**.

The full-rating frequency converter is a back-to-back converter system consisting of two power electronics converters interconnected through a DC-link. These two converters are the generator converter and the grid-side converter. The generator converter absorbs the active power produced by the induction generator and transmits power through a DC-link to the grid-side converter, see **Figure 2.14**. The generator converter is also applied for excitation of the induction generator via its stator terminals. The grid-side converter receives the active power transmitted through the DC-link and delivers it to the power grid, e.g. it balances the DC-link voltage. The grid-side converter is also set to control the power factor or support the grid voltage depending on the chosen control strategy.

Application of smaller gearboxes, reducing the total power losses, variable-speed operation, e.g. better power optimisation, and reactive power controllability using the full capacity of the converter are among the main advantages of this wind turbine concept.

8.1   Induction generator representation and control

An electrical system model of this wind turbine concept is shown in **Figure 8.1.b**. The induction generator feeds into the voltage source induced by the generator converter. The generator converter controls the magnitude and electrical frequency of the voltage induced at the generator terminals, $U_G \angle \alpha$. The electrical frequency of the voltage induced at the generator terminals, f, is coupled to the generator rotor speed, ω_G, through the slip, s, as:

$$\omega_G = f \cdot (1 + s). \tag{8.1}$$

In normal operation, the generator rotor slip, s, may vary in the range of a couple percent. This implies that the fully variable-speed operation of the wind turbine is gained by the fully-variable frequency operation of the induction generator. The frequency may be controlled applying a PI-controller (Xue, 2005) as shown in **Figure 8.2**. Here, P_E is the active power of the generator and $P_{E,REF}$ denotes its reference. The maximum power tracking algorithm is applied to optimise the power reference for gaining the desired electrical frequency at the induction generator terminals, f, see **Figure 2.13**. This algorithm is similar to that applied to optimise the rotor speed of variable-speed wind turbines equipped with DFIG, see **Section 3.1.3**. Rated electrical frequencies may be between 10 and 25 Hz for MW- class wind turbines with converter-connected induction generators.

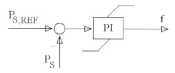

Figure 8.2.   Active power control of converter-connected induction generators.

When the induction generator is connected to the power grid through the AC/DC/AC-converter, the generator and grid do not exchange reactive power. Therefore, the induction generator is excited by the generator converter. To reduce the power rating of the generator converter, fixed capacitors can be applied at the generator terminals as shown in **Figure 8.1.a** and **Figure 2.14**.

Due to the presence of the back-to-back converter system between the induction generator terminals and the power grid, the induction generator may be less affected by a short-circuit fault in the power grid so long as the converter system maintains uninterrupted operation. A reduced third-order model can then be applied to represent the dynamics of the converter-connected induction generators without the loss of accuracy. The model state equations are given in **Section 4.2.2** and model initialisation is described in **Section 4.2.6**.

8.2 DC-link and grid-side converter representation

The DC-link is modelled in the same way as the partial-load frequency converters of DFIG, see **Section 7.4.3**. The DC-link voltage, U_{DC}, is initialised assuming a balance between the injected and extracted active current $J_1 = \dfrac{P_S}{U_{DC}}$ and $J_2 = \dfrac{P_E}{U_{DC}}$. Here P_E denotes the active power supplied to the power grid from the grid-side converter terminals. The initial value of the DC-link voltage is set according to Eq.(7.13) and Eq.(7.14).

The grid-side converter is set to balance the DC-link voltage and to control the reactive power exchanged between the grid-side converter and the power grid. By balancing the DC-link voltage, the active power supplied from the induction generator to the power grid is controlled. The reactive power control arrangement depends on grid requirements. It can be set to optimise the wind turbine power factor, $cos(\varphi) = 1$, or support the grid voltage, keeping it within a desired range. The control system presented in **Figure 7.8.d-e** with independent control of the DC-link voltage and the reactive power may be applied to control the grid-side converter.

When the power grid is subject to a short-circuit fault, the grid voltage drops as does the active power supplied from the grid-side converter to the power grid. If the active power of the induction generator does not reduce fast enough, the DC-link capacitor charges and the DC-link voltage rapidly increases. Excessive over-voltage in the DC-link may interrupt operation of the frequency converter and cause the disconnection of the wind turbine. To prevent this, the frequency converter must have a fault ride-through solution, such as the generator converter may interrupt the active power supply to the DC-link, which stops charging the DC-link capacitor. Additionally, the DC-link may be equipped with a breaking chopper as explained in **Section 7.8.2.1**. The breaking chopper is activated by a converter protective system, discharging the DC-link capacitor and reducing the DC-link voltage during critical operational situations.

8.3 Mechanical construction model

The shaft system transfers the mechanical power of the rotor, P_M, to the generator rotor shaft. The shaft system is represented using the two-mass model by the state equations Eq.(3.46).

The aerodynamic rotor is modelled applying the $C_P(\lambda,\beta)$ characteristics and a model of the pitch control as described in **Section 3.1.4**. Normally, pitch control is arranged using a digital control system with a sampling time in the range of 100 ms. In generic models, this digital control system is commonly represented as a Laplace (continuous) transfer function, as shown in **Figure 3.10**.

8.4 Fault ride-through capability

When a back-to-back converter system contains voltage-sourced converters, operation of the converter system is controlled by Insulated Gate Bipolar Transistor (IGBT) -switches. The IGBT-switches must be protected against thermal and electrical over-loads. This issue is discussed for the case of DFIG in **Section 7.7.1**. A sudden voltage drop during a short-circuit fault may cause excessive current transients in the grid-side converter and also leads to excessive over-voltage in the DC-link, which results in protective blocking of the frequency converter. If the grid-side converter blocks, there is a risk that the wind turbine disconnects from the power grid and stops. Disconnection and stopping of the wind turbine due to a short-circuit fault will be in contrary with the Grid Codes of the national system operators, for example the Grid Code (Energinet.dk, 2004(b)).

To provide a fault-ride-through capability, the grid-side converter may not be allowed permanently to trip at a short-circuit fault. Rather it must maintain uninterrupted operation during grid faults. In critical situations, the grid-side converter may quickly block, but restart already at low-voltage grid operations. When the grid-side converter rides through the fault, the full-rating capacity of this converter can be set to control the reactive power and support grid voltage re-establishment.

When the DC-link voltage increases excessively, the generator converter may interrupt the active power infeed into the DC-link and stop any further charging of the capacitor. This can be gained by two features.

1) The IGBT-switches of the generator converter stop switching and open, then the induction generator is connected to the DC-link capacitor through a diode bridge, **Figure 8.3**. When the residual voltage at the generator terminal (in a single phase) becomes larger than the DC-voltage across the diode bridge, the generator may still inject current into the DC-link. This residual voltage at the generator terminals, U_G, is induced by the current in the rotor circuit, I_R, and can be maintained in seconds. As the current in the rotor circuit decays, the induction generator demagnetises and the residual terminal voltage, U_G, also decays. Tripping of the fixed capacitors at the generator terminals will also contribute to reducing this residual voltage. The current injected into the DC-link stops when the terminal voltage becomes lower than the DC-voltage across the diode bridge. This process may take a few milliseconds.

2) The generator converter continues uninterrupted operation, but changes the control strategy. At excessive values of the DC-link voltage, the reference power $P_{S,REF}$ is immediately reduced or ever set to a negative value. The electrical frequency induced at the generator terminals changes almost immediately to follow the new power reference, see **Figure 8.2**. The generator converter interrupts the power injected into the DC-link or ever absorbs some active power reducing the DC-link voltage. In this control mode, the pitch control will handle the generator speed control.

At critical over-voltage in the DC-link, a breaking chopper may be activated to discharge the DC-link capacitor and bring the DC-link voltage into normal operation. The duration of the breaking chopper action is a few milliseconds due to a risk of thermal over-loading of the chopper itself. Such a breaking chopper consisting of an IGBT-switch and a resistance must be sufficiently reliable.

When the generator converter blocks or sets the power reference to zero, the grid-side converter is operated as a Statcom applying the full-rating capacity of the converter to control reactive power and support grid voltage re-establishment.

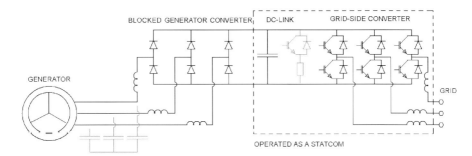

Figure 8.3   Operation of a converter connected induction generator with blocked generator converter during a short-circuit fault.

When the grid voltage has re-established, the generator converter restarts and operation the wind turbine will be back to normal within a few seconds.

8.5   Simulation examples

In this simulation example, a 5 MW wind turbine with a converter connected induction generator is connected to a power system equivalent shown in **Figure 8.4**. The wind turbine is at rated operation and supplies 5 MW to the power grid. The grid disturbance is a transient, 3-phase short-circuit fault applied to the line as marked in **Figure 8.4**. The fault duration is 100 ms. The simulations are performed using the simulation tool Powerfactory^{TM} applying a user-written model of a wind turbine equipped with a full-rating converter-connected induction generator.

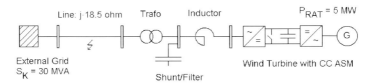

Figure 8.4   The power grid equivalent applied in simulations.

The terminal voltage of the generator and grid-side converter is 0.7 kV, the rated DC-voltage is set to 1.5 kV and the DC-capacitor is chosen to keep DC-voltage ripple below ±1%. The wind turbine must ride through this grid fault as required by the national system operator (Energinet.dk, 2004(b)).

8.5.1. Generator converter blocking and restart

First, a fault ride-through solution applying the generator converter blocking sequence is exam-ined. The simulation results for this control are shown in **Figure 8.5**. When the short-circuit fault is subject to the power grid, the grid voltage and the active power supplied from the converter termi-nals to the power grid drop. This initiates a rapid increase of the DC-link voltage. When the DC-link voltage approaches the value of 1.2 p.u., the generator converter blocks (where 1.0 p.u. corre-sponds to the rated DC-link voltage). The breaking chopper is activated when the DC-link voltage exceeds 1.25 p.u.. These two arrangements keep the converter system operating at a low-voltage.

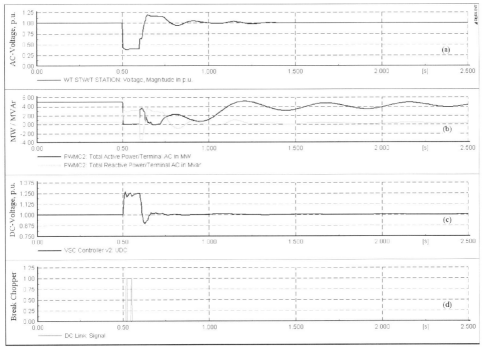

Figure 8.5   Simulated response of a wind turbine with a converter-connected induction generator with the generator converter blocking and restart for a fault-ride-through: **(a)** - grid voltage at the grid-side converter termi-nals, **(b)** - active and reactive power supplied to the grid, **(c)** - DC-link voltage, **(d)** - activation of the break-ing chopper.

During the grid fault, the breaking chopper is activated (twice in this simulation), but in short pe-riods, to avoid any possible thermal over-load. A thermal model of the converter is not part of the examined model and normally will not be applied in investigations of short-term voltage stability. If the simulations predicted an intense use of the breaking chopper to gain the fault-ride-through op-eration of the wind turbine, the results must be interpreted with percussions.

When the grid voltage drops, the grid-side converter is set to control the reactive power and sup-port the grid voltage. This contributes to faster re-establishment of the grid voltage. When the grid fault is removed, the generator converter restarts. Normal operation of the wind turbine re-establishes during a few seconds.

8.5.2. Frequency control

The second solution for the fault ride-through is based on the uninterrupted operation of the generator converter. When the grid voltage drops and the DC-link voltage rapidly increases, the active power reference of the generator converter, $P_{S, REF}$, changes to about zero or a negative value. Applying the control of **Figure 8.2**, the frequency induced by this converter at the generator terminals, f, also changes resulting in almost immediate reduction of the active power injected into the DC-link. **Figure 8.6** presents the simulation results for this fault ride-through solution. When the power reference is set to zero, the DC-link capacitor stops charging. When the power reference is set to a negative value, the generator converter starts absorbing active power from the DC-link and discharges its capacitor. This can be useful if the regular control of the grid-side converter is insufficient to reduce over-voltage in the DC-link during an acceptable period.

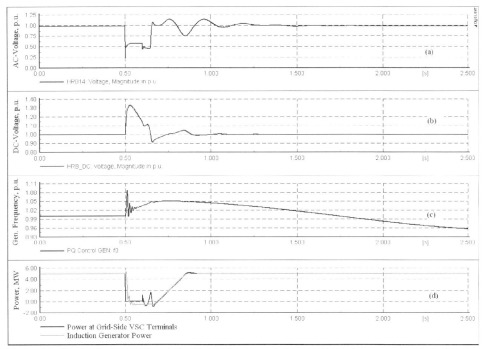

Figure 8.6   Simulated response of a wind turbine with a converter-connected induction generator with the frequency control for fault ride-through to a short-circuit fault: **(a)** - grid voltage at the grid-side converter terminals, **(b)** - DC-link voltage, **(c)** - frequency induced by the generator converter, **(d)** - active power of the induction generator and the grid-side converter.

The grid-side converter operates as a Statcom, where the full converter rating is set to control reactive power and support grid voltage re-establishment. When the fault is removed, the generator converter ramps its power reference back to the pre-fault level. In this simulation, the ramping time is set to 200 ms, but it can be done faster. Note that the active current component of the grid-side converter current is reset when the fault is removed and the grid voltage jumps back to the regular value. This reset sequence might be necessary to avoid excessive active current through the IGBT-switches of the grid-side converter at the time of the fault clearance.

In this simulation, the breaking chopper is not activated (disabled in the model). However, this additional feature could be useful to keep the DC-link voltage below a certain upper limit.

8.6 Summary

Variable-speed wind turbines equipped with full-rating converter-connected induction generators have been described. Such a full-rating frequency converter is a back-to-back frequency converter system consisting of the generator converter and the grid-side converter connected together through a DC-link.

Full-rating frequency converters allow induction generators to operate at fully-variable frequencies. The wind turbines operate at fully variable speeds allowing better power optimisation from incoming wind. In this concept, large mechanical gearboxes can be replaced by a combination of smaller (mechanical) gearboxes and power electronics converters (an electrical "gearbox"). The use of smaller mechanical gearboxes reduces their size, weight and cost. Furthermore, this may also reduce the total power losses in the wind turbine and, then, the wind turbines can start operating at lower wind speeds than those of conventional concepts. However, this concept may require the use of larger induction generators because these operate at lower electrical frequencies.

Induction generators supply active power to the grid through a DC-link with the converter system, but do not exchange reactive power with the grid. Induction generators are excited from generator converters.

The grid-side converter is set to balance the DC-link voltage and control the active power supply from the generator to the power grid. This converter can also be set to control reactive power supply to the power grid and support grid voltage. This controllability is especially useful in remote areas without access to a strong grid and also during grid faults.

The national system operators require the fault-ride-through capability of wind turbines. The fault-ride-through solutions for this wind turbine concept must prevent over-loading and tripping of the full-rating converters. This can be reached by guiding the generator converter through a grid-fault sequence which must prevent excessive over-voltage in the DC-link. For example, the generator converter may interrupt charging the DC-link capacitor when the DC-link voltage approaches a critical level. This can be arranged by either the converter blocking or the frequency control immediately reducing the active power injection into the DC-link. Additionally, the breaking chopper may also be applied to discharge the DC-link capacitor when other arrangements are not sufficient to keep the DC-link voltage below a certain level. When the grid fault is removed, operation of the generator converter returns to normal. During this sequence, the grid-side converter is set to operate as a Statcom and support the grid voltage re-establishment.

The ride-through capability and the ability to control the grid voltage at grid disturbances make this wind turbine concept relevant for application to large (offshore) windfarms.

9   Aggregated models of large windfarms

In investigations of short-term voltage stability, large windfarms may be represented using one of the following options.

1) Detailed models with representation of all wind turbines in the windfarm, all the transformers connecting the wind turbine generators to the internal network of the windfarm. For example, a windfarm containing eighty wind turbines will be represented with eighty wind turbine models, see **Figure 9.1.a**.

2) Aggregated models, which implies that the large windfarm is represented using a single-machine equivalent consisting of a single wind turbine model, see **Figure 9.1.b**, or a multi-machine equivalent consisting of a less number of wind turbine models with re-scaled power ratings, see **Figure 9.1.c-d**. Such aggregated representations can be applied at specific conditions.

The model details of the large windfarm depend on the target of investigations. Detailed models must be applied to investigate if there is any risk of mutual interaction between wind turbines and problems related to the internal network of the windfarm such as power losses, internal faults in the windfarm and protection.

In investigations of short-term voltage stability, focus is on a collective response of the large windfarm to a short-circuit fault in the transmission grid. In this case, an aggregated model of the large windfarm may be applied. The use of an aggregated model is preferred because this simplification reduces the complexity and time of computations.

An aggregated model gives the response of the large windfarm as the integrated whole without distinguishing between each single wind turbine in the windfarm. Therefore, application of the aggregated model may introduce some inaccuracy in the results, and corresponds to the averaging of operational points of the wind turbines and disregards specific operational conditions of wind turbines in the large windfarm. Inaccuracy in the results must be minimised by the accurate representation in the aggregated model using quantitative rules. These aggregation rules must take into account the averaging of the operation points as well as differences in the dynamic response of all the wind turbines.

9.1   General relations

Wind turbines in large windfarms usually have identical generator data and identical mechanical parameters for the shaft systems and aerodynamic rotors. The wind turbines in the windfarm shown in **Figure 9.1** are marked by a pair of indexes (i,j) where the first index refers to a section of the windfarm and goes from 1 to N and the second index refers to a wind turbine in the given section and goes from 1 to M. For the given large windfarm, the indexes are $i =[1,N] =[1,8]$ and $j =[1,M] =[1,10]$, which means eight sections with ten wind turbines in each section.

The (apparent) power capacity, e.g. the MVA-rating, of the aggregated equivalent, $S_{\Sigma\Sigma}$, is the sum of the power capacities of all the wind turbines, $S_{I,J}$.

$$S_{\Sigma,\Sigma} = \sum_{I=1}^{N}\sum_{J=1}^{M} S_{I,J} = N \cdot M \cdot \langle S_{E} \rangle \,. \tag{9.1}$$

Here $\langle S_E \rangle$ denotes the average power capacity of the (indentical) wind turbines in the large windfarm. When applying the reduced model of the large windfarm, this relation is always fulfilled as the windfarm power capacity is the sum of the power capacities of all the wind turbines commissioned in the windfarm.

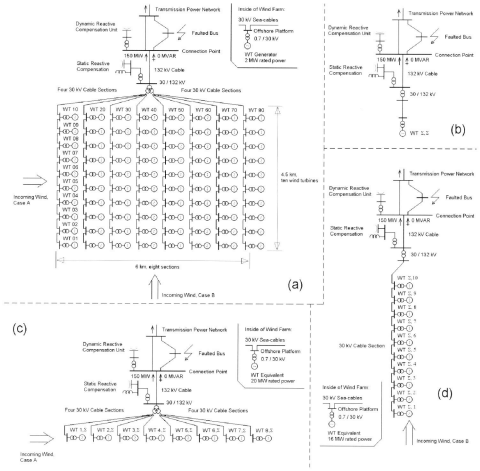

Figure 9.1 Models of a large offshore windfarm: **(a)** - detailed model, **(b)** - a single machine equivalent, **(c)** - a multi-machine equivalent when incoming wind as in Case A and **(d)** - a multi-machine equivalent when incoming wind as in Case B. Reprinted from Ref. (Akhmatov, 2003(b)), Copyright (2003), with permission from the copyright holder.

The active power of the reduced equivalent, $P_{\Sigma,\Sigma}$, is the sum of the active power supplied by all the wind turbines to the power grid, $P_{I,J}$.

$$P_{\Sigma\Sigma} = \sum_{I=1}^{N}\sum_{J=1}^{M} P_{I,J} = N \cdot M \cdot \langle P_E \rangle \ . \tag{9.2}$$

Here $<P_E>$ is the average active power supplied by the wind turbines in the large windfarm. This relation is also always fulfilled as the active power supplied by the windfarm to the power grid is the total active power supplied by all the wind turbines commissioned in the windfarm.

The reactive power of the reduced equivalent, $Q_{\Sigma\Sigma}$, is the sum of the reactive power exchanged by all the wind turbines with the power grid, $Q_{I,J}$.

$$Q_{\Sigma\Sigma} = \sum_{I=1}^{N}\sum_{J=1}^{M} Q_{I,J} = N \cdot M \cdot \langle Q_E \rangle \ . \tag{9.3}$$

Here $<Q_E>$ means the average reactive power exchanged by all the wind turbines in the large windfarm with the power grid. Fulfilment of this relation depends on the specific wind turbine concept and also the specific operation conditions defined for the large windfarm. Additionally, the considerations for the mechanical system must be taken into account.

9.2 Fixed-speed wind turbines

The P-Q-characteristic of the conventional induction generator may be expressed as a parabola with a zero-sloop at no-load operation:

$$Q_{I,J} = Q_{NL} + a \cdot P_{I,J}^2 \ . \tag{9.4}$$

Here, Q_{NL} denotes the reactive absorption of the induction generator at no-load operation and a is a coefficient. When the induction generators of the fixed-speed wind turbines are no-load compensated, the reactive power absorbed by the wind turbines from the power grid becomes:

$$Q_{\Sigma\Sigma} = \sum_{I=1}^{N}\sum_{J=1}^{M} Q_{I,J} - N \cdot M \cdot Q_{NL} = a \cdot \sum_{I=1}^{N}\sum_{J=1}^{M} P_{I,J}^2 \ . \tag{9.5}$$

Aggregation of the large windfarm to the reduced equivalent implies that the average condition for the active power Eq.(9.2) and for the reactive power absorption Eq.(9.5) must be fulfilled simultaneously. Therefore, the following condition must be fulfilled when initialising the large windfarm model.

$$\sum_{I=1}^{N}\sum_{J=1}^{M} P_{I,J}^2 = N \cdot M \cdot \langle P_E^2 \rangle = N \cdot M \cdot \langle P_E \rangle^2 \ . \tag{9.6}$$

This condition is fulfilled when all the wind turbines in the large windfarm supply the same active power to the power grid, $P_{I,J} = P_E$. Fulfilment of this condition is required due to non-linear dependence of the reactive power absorption from the active power of the no-load compensated induction generators Eq.(9.4).

When the induction generators of the fixed-speed wind turbines are fully-compensated, the reactive power absorbed by the large windfarm is zero.

$$Q_{\Sigma\Sigma} = 0 . \tag{9.7}$$

Then, non-linearity of the relation between the reactive power and the active power of the induction generators does not introduce any restrictions on the aggregation of the large windfarm model to a single-machine equivalent. Furthermore, the dynamics of conventional induction generators are given by the sloop of the P-Q-characteristic of the induction generators. For aggregation of the large windfarm model to a single-machine equivalent, the following condition must be fulfilled.

$$\frac{dQ}{dP}_{\Sigma\Sigma} = \sum_{I=1}^{N}\sum_{J=1}^{M}\frac{dQ_{I,J}}{dP_{I,J}} = 2a \cdot \sum_{I=1}^{N}\sum_{J=1}^{M}P_{I,J} = 2a \cdot N \cdot M \cdot \langle P_E \rangle, \tag{9.8}$$

where $dQ_{I,J}/dP_{I,J}$ is the sloop of the P-Q-characteristic of the induction generator of the fixed-speed wind turbine indexed (i,j), and $dQ/dP_{\Sigma\Sigma}$ denotes the sloop of the P-Q-characteristic of the single-machine equivalent of the large windfarm. Obviously, the condition Eq.(9.8) is always fulfilled as this expresses aggregation of the active power of the large windfarm similarly to Eq.(9.2).

Since the shaft systems of fixed-speed wind turbines are characterised by a relatively low stiffness, K_S, the shaft systems are pre-twisted in normal operation and have accumulated a non-negligible amount of potential energy. When the power grid is subject to a short-circuit fault, the shaft systems start relaxation leading to more acceleration of the generator rotors and influencing the grid voltage behaviour. The relation between the shaft system relaxation and the grid voltage behaviour has been explained in **Section 4.3.4**.

The potential energy accumulated in the pre-twisted shaft system of the wind turbine is given by Eq.(3.56). Hence, the total potential energy accumulated in the shaft systems of all the wind turbines in the large windfarm, W_S, is given by the following relation.

$$W_S = \sum_{I=1}^{N}\sum_{J=1}^{M}W_{I,J} = \frac{1}{2K_S}M \cdot N \cdot \langle T_M^2 \rangle, \tag{9.9}$$

where $W_{I,J}$ is the potential energy accumulated in the shaft system of the wind turbine indexed (i,j) and $\langle T_M \rangle$ denotes the average mechanical torque.

For a single-machine equivalent, the total potential energy, $W_{\Sigma\Sigma}$, can be expressed as:

$$W_{\Sigma\Sigma} = \frac{T_{\Sigma\Sigma}^2}{K_S} = \frac{1}{2K_S}M \cdot N \cdot <T_M>^2 . \tag{9.10}$$

Here $T_{\Sigma\Sigma}$ denotes the mechanical torque of the single-machine equivalent. Comparing Eq.(9.9) to Eq.(9.10), it is seen that these two expressions predict the same result when the mechanical torques of all the wind turbines in the large windfarm are the same, i.e. $T_{I,J} = T_M$. In other words, the single machine equivalent can be used to represent the dynamic response of the wind turbines when they work at the same operation point.

Resuming this discussion, a single-machine equivalent can be applied to represent a large windfarm consisting of fixed-speed wind turbines on conditions discussed in (Akhmatov and Knudsen, 2002).

1) When assuming a regular wind distribution over the windfarm area, the wind turbines are at the same operation points. This implies that the mechanical torque, T_M, the active power, P_E, and the reactive power, Q_E, are the same for all the fixed-speed wind turbines in the large windfarm.

2) When the differences between the operation points of the wind turbines are not significant, a single-machine equivalent can also be applied without introducing inaccuracy in the predicted collective response of the large windfarm.

For example, a single-machine equivalent can be applied to investigate the worst case with regard to short-term voltage stability and the fault-ride-through capability which is present when the wind turbines are at rated operation. Then, the applied single-machine equivalent is also at rated operation.

When the wind distribution over the windfarm area is irregular, but follows a pattern, a multi-machine equivalent may be applied instead of a single machine equivalent to reduce the number of applied wind turbine models. As the multi-machine equivalent contains a number of single-machine equivalents, each single-machine equivalent is set-up for a group of the wind turbines having the same operation points. This is illustrated by the following examples.

When the direction of the incoming wind corresponds to Case A in **Figure 9.1.a**, the wind turbines of section 1 shadow the wind turbines of section 2, which again shadow for the wind turbines in section 3, etc. The wind turbines located in the same section experience the same incoming wind and therefore must be at the same operation points.

At the given wind distribution pattern, the windfarm shown in **Figure 9.1.a** may be represented by eight groups, e.g. eight sections consisting of ten wind turbines per a section. Each single group is represented by its single-machine equivalent. Then, the whole windfarm is represented by the multi-machine equivalent consisting of eight single-machine equivalents denoted $WT_{I,\Sigma}$. This multi-machine equivalent is shown in **Figure 9.1.c**.

The rated power capacity of each single-machine equivalent in the multi-machine equivalent is:

$$S_{I,\Sigma} = \sum_{J=1}^{M} S_{I,J} .$$ (9.11)

The active and reactive power of each single-machine equivalent in the multi-machine equivalent are:

$$\begin{cases} P_{I,\Sigma} = \sum_{J=1}^{M} P_{I,J}, \\ Q_{I,\Sigma} = \sum_{J=1}^{M} Q_{I,J}. \end{cases}$$ (9.12)

The relations Eq.(9.11) and Eq.(9.12) are reached applying summation by the wind turbine index $J =[1,M]$ in each section of the large windfarm $I =[1,N]$.

When the direction of the incoming wind is as for Case B in **Figure 9.1.d**, the large windfarm shown in **Figure 9.1.a** can be represented by a multi-machine equivalent consisting of ten single-machine equivalents, $WT_{\Sigma,J}$. The rated power capacity of each single-machine equivalent in the multi-machine equivalent is:

$$S_{\Sigma,J} = \sum_{I=1}^{N} S_{I,J} .$$
(9.13)

The active and reactive power of each single-machine equivalent in the multi-machine equivalent are:

$$\begin{cases} P_{\Sigma,J} = \sum_{I=1}^{N} P_{I,J}, \\ Q_{\Sigma,J} = \sum_{I=1}^{N} Q_{I,J} . \end{cases}$$
(9.14)

These relations are gained using summation by the section index $I =[1,N]$ for each wind turbine group $J =[1,M]$.

Although multi-machine equivalents distinguish between different operation points of the wind turbines and may, therefore, be more accurate than single-machine equivalents, the single machine equivalents are more frequently applied in investigations of short-term voltage stability. As the fixed-speed wind turbines produce the coherent response to a grid fault, the single-machine equivalents become very suitable to investigate the collective response of the large windfarms to such grid faults.

9.3 Variable-speed wind turbines

In the case of variable-speed wind turbines, the same as for the fixed-speed wind turbines considerations shall be made with regard to the aggregation of the rated power capacity Eq.(9.1) and the active power Eq.(9.2). The reactive power is controlled by power electronics converters in the same way in all the wind turbines in the large windfarm.

Since the variable-speed wind turbines produce a coherent response to a grid fault and the power electronics converters are well-tuned (Slootweg, 2003; Akhmatov, 2004(a)), a single-machine equivalent is commonly used in investigations of short-term voltage stability.

A specific concern can be addressed by the representation of the protective systems used in variable-speed wind turbines and their power electronics converters. In terms of the single-machine equivalent, the protective system must block all the power electronics converters at the same time and may also simultaneously disconnect all the wind turbines in the whole windfarm. This predicted action is caused by the assumption in a single-machine equivalent that all the wind turbines operate identically and are initially at the same operational points.

In a real windfarm, the wind turbines may operate at similar, but different, operation points. Therefore, the protective systems of the variable-speed wind turbines may act differently, depending on the operational points of the individual wind turbines. On the other hand, the discrepancy between the collective response of the large windfarm predicted using a single-machine equivalent and the response observed in a real windfarm can be small because the operation points of all the wind turbines in the windfarm are close to each other. **Figure 9.2** compares the measured and the simulated response of the Horns Rev offshore windfarm with 80 variable-speed wind turbines to a short-circuit fault in the transmission power grid.

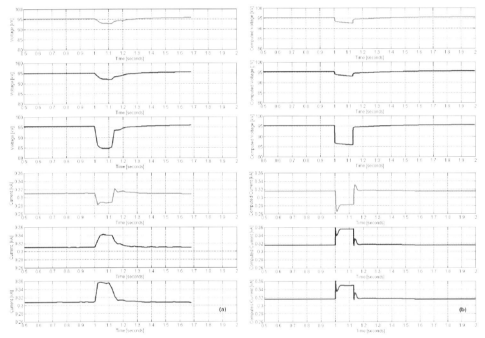

Figure 9.2 Voltage and current in 3 phases in the Danish transmission grid in the vicinity of the Horns Rev connection point (the 150 kV substation Karlsgaarde): (a) - measured with 20 ms sampling and (b) - simulated using a single-machine equivalent. Reprinted from Ref. (Akhmatov and Rasmussen, 2006), Copyright (2006), with permission from Tech-Media.

The simulated response is produced by a single-machine equivalent of the windfarm and with the use of a complete dynamic model of the Danish transmission grid. The response predicted by the single-machine equivalent is in good agreement with the measured response.

9.4 Summary

At certain conditions, windfarms consisting of a large number of wind turbines can be aggregated to single-machine equivalents. The single-machine equivalents represent the whole windfarm as a single wind turbine with a re-scaled rated power capacity. The operation point of the single-machine equivalent (the active and reactive power) is defined by the operation point of the whole windfarm. The main assumption is that the operation points of all the wind turbines in the large windfarm are close to each other or the same (in the ideal case).

Single-machine equivalents are preferred instead of detailed windfarm models in investigations of short-term voltage stability of large power systems because the target is to predict the collective response of the large windfarms to a short-circuit fault in the power grid.

The application of the single-machine equivalent to represent the large windfarm consisting of many wind turbines corresponds to the basic aggregation level applied for the models of all the conventional power plants in a large power grid. This aggregation level means that each large power plant unit is modelled as a single-machine unit, although it may contain several smaller generators.

When the single-machine equivalent is not sufficiently accurate, a multi-machine equivalent can be applied instead, in order to reduce the number of wind turbine models applied in simulations.

10 Retro-fitting with Transient Booster™

Many of the 5,000 Danish wind turbines have been commissioned before the national system operator formulated requirements about fault-ride-through operations. Such wind turbines and wind-farms do not have any fault-ride-through solutions and may trip during a short-circuit fault in the transmission power grid. Therefore the issue of retro-fitting, which means that older wind turbines commissioned without any fault-ride-through solutions can be upgraded to get a fault-ride-through solution, becomes relevant. Retro-fitting may improve the dynamic performance of such older wind turbines and the power grid itself at grid disturbances.

The issues of fault-ride-through capability and reducing the mechanical impact on the wind turbine construction at a short-circuit fault must be treated as two complimentary tasks (Akhmatov, 2003(c); Gertmar et al., 2005). At the end of 2005, ABB A/S announced a retro-fitting solution called Transient Booster™ for improving the fault-ride-through capability of windfarms and reducing mechanical impact on the wind turbine shaft systems (Gertmar et al., 2005). Transient Booster™ can be applied together with fixed-speed wind turbines equipped with conventional induction generators, as well as with variable-speed wind turbines with converter-controlled generators. **Figure 10.1** shows a sketch of Transient Booster™ placed between a wind turbine and the power grid. Transient Booster™ is an electro-mechanic, computer-based sub-system. It consists of a number of switches in each of the three phases and provides an interface between a wind turbine or a windfarm and the electric power grid. When the power grid is in normal operation, the switches are closed and characterised by low electrical resistance in order to minimise power losses.

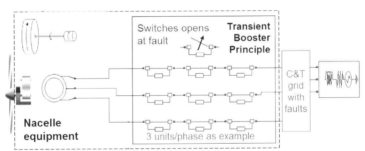

Figure 10.1   Schematic diagram of Transient Booster™ containing electro-mechanic switches and placed between the wind turbine and electric power grid (transmission or distribution). Reprinted from Ref. (Gertmar et al., 2005), Copyright (2005), with permission from ABB A/S.

10.1   Operation at a short-circuit fault

The practical experience of the Danish ABB A/S has shown that unbalanced, 2-phase-to-ground short-circuit faults can be as much as or even more dangerous to induction generators than 3-phase short-circuit faults which occurred near to the generator terminals (Gertmar et al., 2005). At unbalanced, 2-phase-to-ground faults, the electric torque of induction generators affecting the shaft system of the wind turbines can be larger than during balanced, 3-phase short-circuit faults (Gertmar, 1977). Furthermore, such unbalanced faults are more frequent in the transmission systems than 3-phase faults. The long-standing practical experience of the application engineers in the Danish ABB

A/S from wind turbines and their operation has also shown that 2-phase and 2-a-half-phase short-circuit faults may, within parts of a second, develop into 3-phase faults due to ionisation in the low-voltage cable terminations (Gertmar et al., 2005). At 2-phase short-circuit faults, the negative sequence voltage becomes significant and, then, the generator terminal voltage becomes mix of the positive- and the negative-sequence components.

At a short-circuit fault in the transmission system, Transient Booster^{TM} controls switching of its electro-mechanic switches in a way so:

1) The negative-sequence voltage at the wind turbine generator terminals is suppressed;
2) The positive-sequence voltage is maintained close to the rated voltage;
3) Reactive power is directed to reduce demagnetisation of the generator during the grid voltage drop.

Thus, the switches of the phase with a low voltage magnitude are opened and those of the phase with a sufficient voltage magnitude remain closed, see **Figure 10.1**. The switching time is within milliseconds. The switches are almost immediately initiated by, for example over-current registered at the moment of the short-circuit fault occurrence (Gertmar et al, 2005).

10.2 Simulation example

A simulation example to demonstrate the efficiency of Transient Booster^{TM} to maintain the fault-ride-through capability and eliminate excessive mechanical excitation of the wind turbine shaft system is performed with a repetitive, 2-phase short-circuit fault at the grid-side terminals of Transient Booster^{TM}. As the torsion oscillation mode of modern wind turbine shafts is in the range of below 1 to 2 Hz, the duration of the fault is 200 ms and the period of the fault occurrence is 500 ms, chosen to excite the resonant mode of the shaft system. Simulations were performed for a 2 MW fixed-speed wind turbine, described originally by Gertmar et al. (2005) and kindly given to this book by ABB A/S. All simulations are made with omitted protection and constant mechanical torque of the wind turbine rotor. **Figure 10.2** shows the simulation results without the use of Transient Booster^{TM}. As expected, the applied disturbance has excited the undamped torsion oscillation in the shaft system. Already two seconds after the moment of the first short-circuit fault, the generator rotor speed has exceeded the stability limit, which will lead to further excessive overspeeding and disconnection of the wind turbine.

When Transient Booster^{TM} is applied, the reactive power is directed, whereas the negative sequence voltage is suppressed at the generator-side of Transient Booster^{TM}. The positive sequence voltage is kept at a desired level at the generator terminals. Transient Booster^{TM} hinders the repetitive, 2-phase short-circuit fault to excite the resonant mode of the wind turbine shaft system. The generator rotor speed is kept below the speed corresponding to the kip- torque. The wind turbine generator maintains stable operation and the shaft system reduces the mechanical stress during the grid voltage drop in the power grid at the grid-side of Transient Booster^{TM}.

The manufacturer ABB A/S announced that the lost energy through a Transient Booster^{TM} operation is below 10 kWh/event, which is marginal. During any kind of a short-circuit fault, Transient Booster^{TM} can operate with the switches periodically opened or closed or stay with some of the switches opened during a longer period, for example 10 seconds. When the switches are turned off, Transient Booster™ arranged in the low-voltage main circuitry becomes an electric brake (Gertmar et al., 2005).

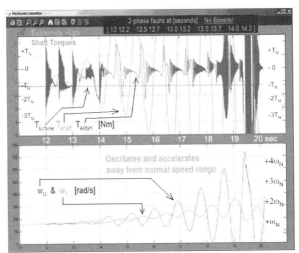

Figure 10.2.   Results for torque and speed without using Transient Booster™. Reprinted from Ref. (Gertmar et al., 2005), Copyright (2005), with permission from ABB A/S.

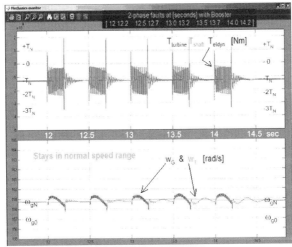

Figure 10.3.   Results for torque and speed using Transient Booster™ - maintaining stable operation. Reprinted from Ref. (Gertmar et al., 2005), Copyright (2005), with permission from ABB A/S.

10.3 Summary

Transient Booster™ from ABB A/S is a retro-fitting solution and cost effective hardware for improvement of the fault-ride-through capability of wind turbines and for reducing shaft system mechanical stress at short-circuit faults. Transient Booster™ contains a number of switches in each phase, which are low-resistance in normal operation of the power grid, but activated at a short-circuit fault. During a short-circuit fault, Transient Booster^{TM} is (i) a transient voltage adaptor, (ii) a transient symmetrizer and (iii) a transient reactive power director.

11 Summarising modelling and stability issues

This book has presented modelling details for wind turbines equipped with induction generators with regard to short-term voltage stability of power grids and the fault-ride-through capability of wind turbines. Although the wind turbines are based on electromechanical power conversion via an induction generator, there is a variety of existing wind turbine concepts applying such induction generators with different kinds of controls.

The conventional concept is fixed-speed, fixed-pitch or active-stall controlled wind turbines equipped with conventional induction generators. The term "fixed-speed" means that the rotor speed is almost constant in normal operation of the power grid and is independent from the operational point of the wind turbine. Rotor speed variation is within the range of the generator rotor slip that may be up to a couple of percent. Conventional induction generators supply active power to the power grid. However, this kind of generator cannot control their magnetisation and are therefore excited from the power grid. In other words, such induction generators absorb reactive power from the power grid. To reduce the reactive absorption from the power grid and improve the power factor of the wind turbines, conventional induction generators are compensated by capacitor banks. Depending on the requirements of the power companies and system operators, wind turbines are either no-load compensated or fully compensated.

When the power grid is subject to a short-circuit fault, the grid voltage drops, the active power is reduced and the rotor accelerates as the mechanical power is not changed during fault durations of up to 100 ms. In conventional induction generators, electrical parameters such as the voltage, active and reactive power are close coupled to mechanical parameters such as generator rotor slip. Therefore, accurate computation of the generator rotor slip is essential for the modelling of fixed-speed wind turbines in investigations of short-term voltage stability.

Short-term voltage stability of fixed-speed wind turbines relates to excessive overspeeding of the wind turbines and is influenced by induction generator and mechanical system data and blade angle control operation during grid faults. Short-term voltage stability is improved when:

1) The rotor resistance is increased and the reactance in stator, in rotor and the magnetising reactance are reduced.
2) The shaft stiffness and the rotor inertia are increased.
3) The blade-angle control is applied under and after the short-circuit fault to prevent excessive overspeeding by reducing the accelerating, mechanical torque of the rotor.
4) Dynamic reactive compensation is applied together with the wind turbines for faster re-establishment of the grid voltage and then for faster increase of the decelerating, electrical torque of the generator.
5) A retro-fitting arrangement is applied.

Fixed-speed wind turbines are a robust concept and may ride through most unbalanced and balanced short-circuit faults in the power grid. This concept is relevant for applications in large (offshore) windfarms. Only in situations when the grid voltage does not recover, or the rotor speed becomes excessive, then wind turbines may disconnect.

Successful development and application of power electronics converters for the control of induction generators has resulted in development of partly- and fully- variable-speed wind turbine concepts. Application of power electronics converters represents a jump in wind technology development leading to wind turbines approaching similar characteristics to conventional power plants equipped with synchronous generators. However, there is still place for improvement.

The manufacturer Vestas Wind Systems produces the OptiSlip® wind turbines which are pitch controlled, partly- variable-speed wind turbines. In partly- variable-speed wind turbines, the power electronics converter is controlled by the Insulated Gate Bipolar Transistor (IGBT) -switches and connected in series to the rotor circuit, which is why this converter is termed a rotor converter. Operation of the rotor converter corresponds to the addition of a dynamic rotor resistance to the rotor circuit impedance. This allows controlled, dynamic variation of the generator rotor slip in the range of 1% to 10% when the wind turbine is at rated operation. In this concept, pitch control is used to optimise the power output of the rotor and rotor converter control is applied to reduce the active power fluctuations - the flicker. Such power fluctuations are transformed into dynamic slip variations and absorbed in the dynamic rotor resistance. In this way, such power fluctuations do not reach the power grid, so improving the power quality. The rotor converter is not designed to control the reactive power of the induction generator. The generator is still excited from the power grid and compensated by a capacitor bank as in the case of fixed-speed wind turbines.

Since the induction generator has increased rotor resistance and pitch control is applied to prevent excessive overspeeding of the wind turbine, this improves the short-term voltage stability.

The IGBT-switches of the rotor converter are sensitive to thermal and electrical overloads. When the power grid is subject to a short-circuit fault, the grid voltage drops. A sudden voltage drop may excite excessive current transients in the rotor circuit which affects the operation of the rotor converter. At excessive transients in the rotor current, the rotor converter blocks so that the IGBT switches stop switching and open. The converter blocking leads to disconnection of the wind turbines, unless the fault ride-through solution with blocking and restart of the rotor converter is applied.

When operating with a blocked rotor converter, the rotor circuit "sees" the total external resistance of the converter. This contributes to efficient and fast dampening of excessive transients in the rotor current. When the grid fault is well-cleared and the grid voltage has re-established, the rotor converter automatically restarts and the wind turbine rides through the grid fault.

In pitch controlled, variable-speed wind turbines equipped with DFIG, the stator feeds directly into the power grid whereas the partial-load frequency converter provides connection between the rotor circuit of the induction generator and the power grid. The frequency converter is required in this concept as it connects the power grid operating at an almost fixed frequency with the rotor circuit operating at a variable frequency. The frequency converter consists of the rotor converter and the grid-side converter connected together through a DC-link. This is often called the back-to-back converter system.

The rotor converter induces an external voltage in the rotor circuit. The frequency of the induced voltage relates to the rotor speed. Then, the rotor converter controls the rotor speed and optimises the power output of the wind turbine. The generator rotor slip may vary in the range from -50 to +15% (and dynamically up to 30%) which corresponds to fully variable-speed operation. The main contribution to the power output optimisation is reached using variable-speed operation of the rotor. Pitch control is applied for better optimisation of the power output of the rotor.

The rotor converter control is arranged using the cascade control and provides independent control of the active and reactive power of the generator. The generator is excited through the rotor circuit, applying the reactive power control of the rotor converter. Then, the generator can operate with optimised power factor (at zero reactive power exchange with the power grid) or be set to support the grid voltage and exchange a desired amount of reactive power with the grid. For example, the manufacturer GE Wind Energy applies a patented solution termed WindVAR^{TM} to actively support the grid voltage via converter control when wind turbines are erected in remote locations. The reactive power control is however limited by the converter rating.

The grid-side converter is set to balance the DC-link voltage and to control the reactive power at its terminals with the use of cascade control. Normally, the reactive power is kept zero at the grid-side converter terminals to improve the power factor and to minimise losses.

The partial-load frequency converter provides generator control, but it is also the most sensitive part of the wind turbine with regard to short-circuit faults in the power grid. The frequency converter is controlled by the IGBT-switches which are sensitive to thermal and electrical overloads. At balanced, 3-phase short-circuit faults, excessive transients may be present in the rotor circuit and excessive voltage fluctuations may occur in the DC-link. Such excessive transients in the rotor circuit and excessive fluctuations in the DC-link voltage may cause the converter blocking. Furthermore, the converter control may be sensitive to unbalanced grid faults such as phase-to-ground or phase-to-phase short-circuit faults. Such unbalanced grid faults may also initiate converter blocking. When blocking, the IGBT-switches stop switching and the frequency converter may trip. The characteristic blocking time is in the range of only milliseconds. When the frequency converter has blocked, the controllability of the DFIG is lost and the wind turbine may trip from the power grid.

The main issue for this wind turbine concept is maintaining the fault-ride-through operation. This can be achieved in several ways. Crowbar protection is among the well-described control solutions providing the fault-ride-through operation of the DFIG based wind turbines. In terms of this solution, the rotor converter may block when affected by excessive transients in the rotor current and automatically restart when the grid operation has re-established to normal and the current transients are well-damped. When the rotor converter has blocked, this converter stops switching and trips and the rotor circuit is short-circuited through a crowbar, e.g. an external resistance. So long as crowbar protection is active, the generator operates as a conventional induction generator. The generator supplies the active power to the grid. The pitch control is set to keep the rotor speed at the desired level and prevent excessive overspeeding when the grid voltage is under re-establishment. As the controllability of the DFIG is lost, the generator is excited (absorbs the reactive power) from the power grid.

During operation with crowbar protection, the grid-side converter maintains uninterrupted operation, e.g. rides through the fault. This converter is set to balance the DC-link voltage and to control the reactive power at the converter mains. Operation of this converter corresponds to a Statcom which is why this converter may be set to support the grid voltage near to of the wind turbine terminals. The controllability of the grid-side converter with regard to the reactive power and support of the grid voltage is restricted by the converter current rating.

The fault-ride-through solution applying the crowbar protection does not require any sufficient increase of the rotor converter rating. The crowbar must be reliable as it provides the fault-ride-through capability for variable-speed wind turbines.

The fault-ride-through solution with current limitation and control may require increasing the power rating of the rotor converter. This solution is based on (almost) uninterrupted operation of the rotor converter during short-circuit faults in the power grid. When the grid fault is registered and the rotor current transients approach the critical level, the rotor converter blocks, but can restart at low-voltage operation of the grid. The rotor converter changes the switching pattern to limit the current going through the IGBT-switches and guides the active and reactive power of the generator within the current rating of the rotor converter. Since the rotor converter maintains uninterrupted operation, the DFIG can be set to control the reactive power and support the re-establishment of the grid voltage.

The grid-side converter maintains uninterrupted operation, balances the DC-link voltage to provide the power exchange between the rotor circuit and the power grid and controls the reactive power at the converter terminals. This converter may also be set to support the grid voltage requiring coordination with the reactive power control of the rotor converter. This is required to avoid conflicts between control systems trying to control the same parameter - the grid voltage near to the wind turbine terminals.

DFIG based wind turbines equipped with fault-ride-through will for "pedagogical reasons" also be equipped with crowbar protection. The crowbar is activated as "the last defence line" to ride through the grid fault if the ordinary fault-ride-through solution is insufficient or fails.

Wind turbine manufacturers have developed specific fault-ride-through solutions for their DFIG based wind turbines. For example, the manufacturer GE Wind Energy offers the LVRT^{TM} solution for large (offshore) windfarms. The LVRT^{TM} solution allows real fault-ride-through operation with supply of reactive power to the grid during faults and with the support of grid voltage re-establishment.

Pitch-controlled, variable-speed wind turbines equipped with full-rating converters and induction generators are a relatively new concept for MW-class wind turbines. First commercial wind turbines of this concept with ratings above 2 MW are announced in summer 2005 by Siemens. There are not much published information on the design and control of these wind turbines yet.

This concept requires a full-rating frequency converter connecting the induction generator with a short-circuited rotor circuit to the power grid. The full-rating frequency converter is a back-to-back converter system with a DC-link between the two converters. Presence of the DC-link between the induction generator terminals and the power grid operating at a fixed frequency allows fully variable-frequency operation of the generator. The induction generator supplies the active power to the power grid through the DC-link, but it is magnetised by the generator converter. The generator converter induces the generator terminal voltage to a required magnitude and electrical frequency. The induced electrical frequency relates to the optimised rotor speed. Then, the rotor produces maximum mechanical power. In this way, the generator converter controls the rotor speed within a similar range as for DFIG based wind turbines.

The grid-side converter is set to balance the DC-link voltage and control the reactive power exchanged with the power grid. Such wind turbines may be set either to operate at optimised power factor or control the grid voltage (when installed in remote locations).

The use of the full-rating frequency converters increases costs of the power electronics when compared to DFIG based wind turbines. However, this concept may have advantages that make it competitive and attractive.

1) The mechanical gear ratio and so the size, weight and cost of the gearbox applied in the variable-speed wind turbine is reduced because the frequency converter contributes to the frequency conversion between the mechanical frequency of the rotor and the electrical frequency of the power grid.

2) When a smaller gearbox is applied together with the full-scale frequency converter instead of a large gearbox, the total losses in the construction may be reduced. As the no-load losses are smaller, the wind turbines may start operation in a lower wind speed when compared to conventional concepts.

However, operation at a lower frequency may lead to the generator size increasing. The grid-side converter is controlled by the IGBT-switches and may block due to excessive over-current or excessive over-voltage in the DC-link appearing at a grid disturbance. To provide the fault-ride-through operation of such converter-connected induction generators, the grid side converter must either ride through the grid fault or be able to automatically restart at low-voltage operations of the power grid (during grid faults) with maintaining all the vital control functions.

Fault-ride-through solutions can be based on the immediate blocking of the generator converter or on immediate reduction of the active power supply from the generator converter to the DC-link at excessive DC-link voltage. This reduces excessive over-voltage in the DC-link and helps the grid-side converter to ride through the fault and handle the reactive power control. The pitch control keeps the rotor speed within a desired range and prevents overspeeding of the rotor. Additionally, a breaking chopper may be incorporated in the DC-link and activated to discharge the DC-link capacitor in situations where the DC-link voltage approaches a critical level. When the DC-link voltage is re-established to a normal range, the generator converter automatically restarts. The generator converter excites the induction generator and ramps up the active power generation to normal.

All the described wind turbine concepts with utilisation of induction generators and their control are able to maintain the fault-ride-through operation. The specific solution details may depend on the chosen wind turbine concept and are also manufacturer-dependent.

Dynamic models for wind turbines to be used for investigations of short-term voltage stability may contain the following sub-models.

1) The transient fifth-order or the reduced third-order model of induction generators. The transient, fifth-order model will be preferred because it represents the fundamental-frequency transients in the machine current affecting the generator rotor speed behaviour of conventional induction generators at a short-circuit fault. This is relevant for the accurate prediction of the rotor overspeeding. Furthermore, excessive transients in the rotor current may cause converter blocking of the converter-controlled generators. Accurate prediction of the converter behaviour and controllability at low-voltage operations is very relevant for stability investigations. When the transient, fifth-order model is not available in a given simulation tool, the reduced, third-order model can be applied under careful guidance of the wind turbine manufacturer regarding representation of the protective system.

2) The shaft system model used for all wind turbine concepts is the two-mass model. This model presents shaft torsion oscillations and is required for the accurate prediction of the rotor speed. The lumped-mass model will predict lower overspeeding and neglect power and voltage fluctuations at grid disturbances.

3) The aerodynamic model of the rotor to compute the mechanical power and the torque supplied to the shaft system. Commonly, rotor aerodynamics are represented using the power coefficient characteristics, $C_P(\lambda,\beta)$. This approach predicts sufficiently accurate behaviour of the rotor torque when active-stall control is applied. When pitch control is applied, this approach should not be used for computations with fast pitching because such fast pitching may result in notable overshoots in the rotor torque, and which is not presented by the power coefficient characteristics' approach.

4) Blade-angle control is part of the mechanical control applied to optimise the rotor operation in low wind and to keep the rotor power at the rated value in strong wind. The most of fixed-speed wind turbines are with active-stall blade-angle control operating in the negative range of the pitch angle. Variable-speed wind turbines are equipped with pitch control that operates in the positive range of the pitch angle. Active-stall is more sensitive to the pitch angle change than pitch control which is why the pitch rates of the active-stall are normally smaller than the pitch rates of similar, pitch controlled wind turbines. The blade-angle control model consists of the regular control system model setting the pitch reference and the servo model adjusting the factual pitch angle to the reference. Additionally, the blade-angle control may be accessed by the signal from the external system that may order the wind turbines to change the operation point. This functionality must be represented in the blade-angle control model, when required.

5) The converter model is part of the electrical control system applied onto the induction generators of variable-speed wind turbines. The converter model must have an interface model to exchange the signals between the converter and the generator models, the control model, the protective system model and the model of the fault ride-through solution. The individual representation of the IGBT-switches is not required as the characteristic switching time is much smaller than the characteristic period of the processes relevant for short-term voltage stability.

6) The protective system model is relevant to predict disconnection of the wind turbines leading to the active power loss in the grid or loss of the controllability when the converter blocks. The protective system monitors the relevant parameters in the generator, the converter and the grid and orders the generator to trip or the converter to block when at least one of the monitored parameters exceeds its relay setting.

The models must be validated if they are to make sense for simulations. The user-written models implemented in a chosen simulation tool must not be in conflict with the network solution algorithms (or with the other models) of the simulation tool.

References

Akhmatov, V., Knudsen, H. (1999). Modelling of windmill induction generators in dynamic simulation programs, *In International IEEE Power Tech. Conference*, Budapest, Hungary, Paper BPT99-243-12.

Akhmatov, V., Nielsen, A.H., Knudsen, H. (2000(a)). Electromechanical interaction and stability of power grids with windmills, *In IAESTED International Conference on Power and Energy Systems*, Marbella, Spain, pp. 398-405.

Akhmatov, V., Knudsen, H., Nielsen, A.H. (2000(b)). Advanced simulation of windmills in the electric power supply, *Electrical Power and Energy Systems*, vol. 22, no. 6, pp. 421-434.

Akhmatov, V., Knudsen, H., Bruntt, M., Nielsen, A.H., Pedersen, J.K., Poulsen, N.K. (2000(c)). A dynamic stability limit of grid-connected induction generators, *In IAESTED International Conference on Power and Energy Systems*, Marbella, Spain, pp. 235-244.

Akhmatov, V., Knudsen, H., Nielsen, A.H., Pedersen, J.K., Poulsen, N.K. (2001). Short-term stability of large-scale wind farms, *In European Wind Energy Conference EWEC-2001*, Copenhagen, Denmark, Paper PG 3.56.

Akhmatov, V. (2001). Note concerning the mutual effects of grid and wind turbine voltage stability control, *Wind Engineering*, vol. 25, no. 6, pp. 367 –371.

Akhmatov, V., Knudsen, H. (2002). An aggregate model of grid-connected, large-scale, offshore wind farm for power stability investigations – Importance of windmill mechanical system, *Electrical Power and Energy Systems*, vol. 24, no. 9, pp. 709 –717.

Akhmatov, V. (2002(a)). Variable speed wind turbines with doubly-fed induction generators, Part I: Modelling in dynamic simulation tools, *Wind Engineering*, vol. 26, no. 2, pp. 85-107.

Akhmatov, V. (2002(b)). Variable speed wind turbines with doubly-fed induction generators, Part II: Power system stability, *Wind Engineering*, vol. 26, no. 3, pp. 171-188.

Akhmatov, V., Knudsen, H., Nielsen, A.H., Pedersen, J.K., Poulsen, N.K. (2003(a)). Modelling and transient stability of large wind farms, *Electrical Power and Energy Systems*, vol. 25, no. 1, pp. 123-144.

Akhmatov, V. (2003(a)). Voltage stability of large power networks with a large amount of wind power, *In Fourth International Workshop on Large-Scale Integration of Wind Power and Transmission Networks for Offshore Wind farms*, Billund, Denmark, 10 p.

Akhmatov, V. (2003(b)). *Analysis of Dynamic Behaviour of Electric Power Systems with Large Amount of Wind Power*, PhD dissertation, Technical University of Denmark, Kgs. Lyngby, Denmark, 254 p.

Akhmatov, V. (2003(c)). Mechanical excitation of electricity producing wind turbines at grid faults, *Wind Engineering*, vol. 27, no. 4, pp. 257-272.

Akhmatov, V. (2003(d)). Variable-speed wind turbines with doubly-fed induction generators. Part IV: Uninterrupted operation features at grid faults with converter control coordination, *Wind Engineering*, vol. 27, no. 6, pp. 519-529.

Akhmatov, V. (2003(e)). Variable-speed wind turbines with doubly-fed induction generators. Part III: Model with the back-to-back converters, *Wind Engineering*, vol. 27, no. 2, pp. 79 -91.

Akhmatov, V. (2004(a)). An aggregated model of a large wind farm with variable-speed wind turbines equipped with doubly-fed induction generators, *Wind Engineering*, vol. 28, no. 4, pp. 479-486.

Akhmatov, V., Nielsen, A.H. (2005). Fixed-speed active-stall wind turbines in offshore applications, *European Transactions on Electrical Power*, vol. 15, no. 1, pp. 1-12.

Akhmatov, V. (2005(a)). Full-load converter connected asynchronous generators for MW class wind turbines, *Wind Engineering*, vol. 29, no. 4, pp. 341-351.

Akhmatov, V. (2005(b)). *Full-scale Verification of Dynamic Wind Turbine Models*, Chapter 27 in Ackermann, T. (Editor), "Wind Power in Power Systems", John Willey & Sons, Ltd., London, U.K., 2005, 24 p.

Akhmatov, V., Rasmussen, C. (2006). Large offshore windfarms: Investigations and modelling for grid-connection to the Danish power system, Elteknik, no. 1, pp. 16-18, in Danish.

McArdle, J. (2004). Dynamic modelling of wind turbine generators and the impact on small lightly interconnected grids, *Wind Engineering*, vol. 28, no. 1, pp. 57-74.

Barocio, E., Messina, A.R. (2003). Normal form analysis of stressed power systems: Incorporation of SVC models, *Electrical Power and Energy Systems*, vol. 25, no. 1, pp. 79-90.

Bolik, S.M. (2003). Grid requirements challenges for wind turbines, *In Fourth International Workshop on Large-Scale Integration of Wind Power and Transmission Networks for Offshore Wind farms*, Billund, Denmark, 6 p.

Bolik, S.M. (2004). *Modelling and Analysis of Variable Speed Wind Turbines with Induction Generator during Grid Fault*, PhD dissertation, Aalborg University Centre, Kgs. Lyngby, Denmark, 198 p.

Bruntt, M., Havsager, J., Knudsen, H. (1999). Incorporation of wind power in the East Danish power system, *In International IEEE Power Tech. Conference*, Budapest, Hungary, Paper BPT99-202-50.

Cadirci, I., Ermis, M. (1992). Double-output induction generator operating at sub-synchronous and super-synchronous speeds: steady-state performance optimisation and wind-energy recovery, *IEE Proceedings-B, Electric Power Applications*, vol. 139, no. 5, pp. 429-442.

Chaviaropoulos, P.K. (1996). Development of a state-of-the art aeroelastic simulator for horizontal axis wind turbines, Part 1: Structural aspects, *Wind Engineering*, vol. 20, no. 6, pp. 405-421.

Cigré. (1999). *Static Synchronous Compensator (Statcom)*, Edited by Erinmez, I.A. and Foss, A.M., Cigré Working Group 14.19, Report 144.

Diez, A., Garcia, D.F., Cancelas, J.A., López, H. (1989). Adaptive control of an asynchronous eolic generator using the field-oriented technique, *In Electrotechnical Conference "Integrating Research, Industry and Education in Energy and Communication Enginnering", MELECON'89, Mediterranean*, pp. 99-103.

Eek, J., Pedersen, K.O.H., Sobrink, K.H. (2004). Development and testing of ride-through capability solutions for wind turbine with doubly-fed induction generator using VSC transmission, *In Cigré Session 2004*, Paris, France, Paper B4-302, 8 p.

Eel-Hwan Kim, Sung-Bo Oh, Yong-Hyun Kim, Chang-Su Kim (2000). Power control of a doubly fed induction machine without rotational transducers, *In 3rd International Conference on Power Electronics and Motion Control, PIEMC-2000*, vol. 2, pp. 951-955.

Eltra (2000). *Specifications for Connecting Wind Farms to the Transmission Network*, 2nd Ed., ELTRA Transmission System Planning, Energinet.dk, Denmark: ELT1999-411a.

Energinet.dk. (2004(a)). *Wind Turbines Connected to Grids with Voltages below 100 kV - Technical regulations for the Properties and the Control of Wind Turbines*, Energinet.dk, Transmission System Operator of Denmark, Technical Regulations TF 3.2.6, 41 p.

Energinet.dk. (2004(b)). *Wind Turbines Connected to Grids with Voltages above 100 kV - Technical regulations for the Properties and the Control of Wind Turbines*, Eltra, Transmission System Operator of Western Denmark, Technical Regulations TF 3.2.5, 35 p.

Eriksen, P.B., Orths, A.G., Akhmatov, V. (2006(a)). Integrating dispersed generation into the Danish power system - Present situation and future prospects, *In IEEE 2006 General Meeting, Montreal, Canada, Panel Session: Impact of Dispersed and Renewable Generation on System Structure Including Impact of Enlarged Community and Energy Development, Power Generation, International Interconnections, Transmission and Distribution*, pp. 7.

Eriksen, P.B., Akhmatov, V., Orths, A.G. (2006(b)). Managing 23%: Grid-integration of wind power in Denmark, *Renewable Energy World*, vol. 9, no. 4 (July-August), pp. 214-227.

Eriksson, K., Halvarsson, P., Wensky, D., Hausler, M. (2003). System approach on designing an offshore windpower grid connection, *In Fourth International Workshop on Large-Scale Integration of Wind Power and Transmission Networks for Offshore Wind farms*, Billund, Denmark, 11 p.

Feijóo, A., Cidras, J., Carrillo, C. (2000). A third order model for the doubly-fed induction machine, *Electric Power Systems Research*, vol. 56, pp. 121-127.

Fortmann, J. (2003). Validation of DFIG model using 1.5 MW turbine for the analysis of its behaviour during voltage drops in the 110 kV grid, *In Fourth International Workshop on Large-Scale Integration of Wind Power and Transmission Networks for Offshore Wind farms*, Billund, Denmark, 5 p.

Freris, L. (1990). *Wind Energy Conversion Systems*, Prentice Hall, New York, 388 p, in Chapter 4.

Gertmar, L. (1977). *Calculation of maximal currents and torques at single-, two- and three-phase short-circuits of asynchronous generator*, Chalmers Univ. of Technology, Dept. of Electrical Machinery, Gothenburg, 1977. -

Gertmar, L., Christensen H.C., Nielsen, E.K., Wraae, L.E. (2005). New method and hardware for grid-connection of wind turbines and parks, *Proc. Copenhagen Offshore Wind Conference and Exhibition*, Copenhagen, Denmark, 9 p.

Gjengedal, T., Gjerde, J.O., Sporild, R. (1999). Assessing of benefits of adjustable speed hydro machines, *In IEEE International Power Tech 99 Conference*, Budapest, Hungary, Paper BPT99-439-21.

Hansen, A., Bindner, H. (1999). *Combined Variable Speed/ Variable Pitch Controlled 3-bladed Wind Turbine: Control Strategies*, Risø National Laboratory, Roskilde, Denmark, 47 p.: Risø-R-1071(DA), in Danish.

Hansen M.O.L. (2000). *Aerodynamics of Wind Turbines: Rotors, Loads and Structure*, James & James, London, U.K., 144 p.

Hansen, A.D., Iov, F., Blaabjerg, F, Hansen, L.H. (2004(a)). Review of contemporary wind turbine concepts and their market penetration, *Wind Engineering*, vol. 28, no. 3, pp. 247-263.

Hansen, A.D., Sørensen, P., Iov, F., Blaabjerg, F. (2004(b)). Control of variable speed wind turbines with doubly-fed induction generators, *Wind Engineering*, vol. 28, no. 4, pp. 411-432.

Hansen, M.H., Hansen, A.D., Larsen, T.J., Oye, S., Sørensen, P.E., Fuglsang, P. (2005). *Control design for a pitch-regulated, variable speed wind turbine*, Risø National Laboratory, Roskilde, Denmark, 84 p.: Risø-R-1500(EN).

Heier, S. (1996). *Windkraftanlagen im Netzbetrieb*, 2.Aufl., Teubner, Stuttgart, Germany, 396 p.

Hinrichsen, E.N., Nolan, P.J. (1982). Dynamic stability of wind turbine generators, *IEEE Transactions on Power Apparatus Systems*, vol. PAS-101, no. 8, pp. 2640-2648.

Hinrichsen, E.N. (1984). Controls for variable pitch wind turbine generators, *IEEE Transactions on Power Apparatus and Systems*, vol. PAS-103, no. 4, pp. 886-892.

Holdsworth, L., Jenkins, N., Strbac, G. (2001). Electrical stability of large, offshore wind farms, *In IEE AC-DC Power Transmission Conference*, Publication No. 485, pp. 156-161.

Holdsworth, L., Charalambous, I., Ekanayake, J.B., Jenkins, N. (2004). Power system fault ride through capabilities of induction generator based wind turbines, *Wind Engineering*, vol. 28, no. 4, pp. 399-409.

Hogdahl, M., Nielsen, J.G. (2005). Modelling of the Vestas V80 VCS wind turbine with low voltage ride-through, *Proc. Fifth Int. Workshop on Large-Scale Integration of Wind Power and Transmission Networks for Offshore Wind Farms*, Glasgow, Scotland, pp. 292-298.

Ibrahim, E.S. (1997). Digital simulation of electromagnetic transients, *Electric Power Systems Research*, vol. 41, pp. 19-27.

Jones, B.L., Brown, J.E. (1984). Electrical variable speed drives, *IEE Proceedings-A*, vol. 131, no. 7, pp. 516-558.

Krause, P.C., Wasynczuk, O., Sudoff, S.D. (1995). *Analysis of Electric Machinery*, IEEE Power Engineering Society, N.Y., U.S.A., 1995, in Chapter 3.

Kristoffersen, J.R., Christiansen, P. (2003). Horns Rev offshore wind farm: its main controller and remote control system, *Wind Engineering*, vol. 27, no. 5, pp. 351-360.

Krüger, T., Andresen, B. (2001). Vestas OptiSpeed[TM] – Advanced control strategy for variable speed wind turbines, *In European Wind Energy Conference EWEC-2001*, Copenhagen, Denmark, pp. 983-986.

Kundur, P. (1994). *Power System Stability and Control*, EPRI, McGraw-Hill, New York, 1176 p.

Ledesma, P., Usaola, J., Rodriguez, J.L., Burgos, J.C. (1999). Comparison between control systems in a doubly fed induction generator connected to an electric grid, *In European Wind Energy Conference EWEC-1999*, Madrid, Spain, 4 p.

Leith, D.J., Leithead, W.E. (1997). Implementation of wind turbine controllers, *International Journal of Control*, 1997, vol. 66, no. 3, pp. 349-380.

Lindholm, M. (2004). *Doubly Fed Drives for Variable Speed Wind Turbines, A 40 kW Laboratory Setup*, PhD dissertation, Technical University of Denmark, Kgs. Lyngby, Denmark, 154 p.

Miller, N.W. (2003). Power system dynamic performance improvements from advanced power control of wind turbine generators, *In Fourth International Workshop on Large-Scale Integration of Wind Power and Transmission Networks for Offshore Wind farms*, Billund, Denmark, 6 p.

Müller, S., Deicke, M., De Doncker, R. W. (2000). Adjustable speed generators for wind turbines based on doubly-fed induction machines and 4-quadrant IGBT converters linked to the rotor, *In IEEE Industry Applications Conference*, Rome, Italy, vol. 4, pp. 2249-2254.

Neris, A.N., Vovos, N.A., Giannakopoulos, G.B. (1996). Dynamics and control system design of an autonomous wind turbine, *In Third IEEE International Conference on Electronics, Circuits and Systems ICECS'96*, Rhodes, Greece, vol.2, pp. 1017-1020.

Noroozian, M., Knudsen, H., Bruntt, M. (2000). Improving a wind farm performance by reactive power compensation, *In IAESTED International Conference on Power and Energy Systems*, Marbella, Spain, pp. 437-442.

Næss, B.I., Undeland, T.M., Gjengedal, T. (2002). Methods for reduction of voltage unbalance in weak grids connected to wind plants, *In IEEE/Cigré Workshop on Wind Power and the Impacts on Power Systems*, Oslo, Norway, 6 p.

Papathanassiou, S.A., Papadopoulos, M.P. (1999). Dynamic behaviour of variable speed wind turbines under stochastic wind, *IEEE Transactions on Energy Conversion*, vol. 14, no. 4, pp. 1617-1623.

Pedersen, J.K., Akke, M., Poulsen, N.K., Pedersen, K.O.H. (2000). Analysis of wind farm islanding experiment, *IEEE Transactions on Energy Conversion*, vol. 15, no. 1, pp. 110-115.

Pedersen, J.K., Pedersen, K.O.H., Poulsen, N.K., Akhmatov, V., Nielsen, A.H. (2003), Contribution to a dynamic wind turbine model validation from a wind farm islanding experiment, *Electric Power Systems Research*, vol. 64, no. 1, pp. 41-51

Pena, R., Clare, J.C., Asher, G.M. (1996). Doubly-fed induction generator using back-to-back PWM converters and its application to variable-speed wind-energy generation, *IEE Proceedings-B, Electric Power Applications*, vol. 143, no. 3, pp. 231-241.

Pena, R.S., Cardenas, R.J., Asher, G.M., Clare, J.C. (2000). Vector controlled induction machines for stand-alone wind energy applications, *In IEEE Industry Applications Conference*, Rome, Italy, vol. 3, pp. 1409-1415.

Petersson, A. (2003). *Analysis, modelling and control of doubly-fed induction generators for wind turbines*, Chalmers University of Technology, Göteborg, Sweden, 122 p.: Technical Report No. 464L.

Petersson, A. (2005). *Analysis, modelling and control of doubly-fed induction generators for wind turbines*, Dept. of Energy and Environment, Chalmers University of Technology, Göteborg, Sweden, 166 p.

Pretlove, A.J., Mayer, R. (1994). Rotor size and mass – The dilemma for designers of wind turbine generating systems, *Wind Engineering*, vol. 18, no. 6, pp. 317-328.

Pöller, M. (2003). Doubly-fed induction machine models for stability assessment of wind farms, *In International IEEE Power Tech. Conference*, Bologna, Italy, 6 p.

Raben, N., Donovan, M.H., Jørgensen, E., Thisted, J., Akhmatov, V. (2003). Grid tripping and re-connection: Full-scale experimental validation of a dynamic wind turbine model, *Wind Engineering*, vol. 27, no. 2, pp. 205 -13.

Rasmussen, C., Jørgensen, P., Havsager, J., Nielsen, B., Andersen N. (2005). Improving voltage quality in Eastern Denmark with a dynamic phase compensator, *In Fifth International Workshop on Large-Scale Integration of Wind Power and Transmission Networks for Offshore Wind Farms*, Glasgow, Scotland, pp. 387-392.

Røstøen, H.O., Undeland, T.M., Gjengedal, T. (2002). Doubly-fed induction generator in a wind turbine, *In IEEE/Cigré Workshop on Wind Power and the Impacts on Power Systems*, Oslo, Norway, 6 p.

Salman, S.K., Teo, A.L.J. (2003). Windmill modelling consideration and factors influencing the stability of a grid-connected wind power based embedded generator, *IEEE Trans. on Power Systems*, vol. 18, no. 2, pp. 793-802.

Santos, J.S. (2002). *Request for Proposal: ERCOT Wind Generation Models and Model Validation*, ERCOT, Texas, U.S.A. 22 p.

Schauder, C., Metha, H. (1993). Vector analysis and control of advanced static VAR compensators, *IEE Proceedings-C*, vol. 140, no. 4, pp. 299-306.

Slootweg, J.G., Polinder, H., Kling, W.L. (2001). Initialisation of wind turbine models in power dynamics simulations, *In 2001 IEEE Porto Power Tech Conference*, Porto, Portugal, 6 p.

Slootweg, J.G. (2003). *Wind Power Modelling and Impact on Power System Dynamics*, PhD thesis, Delft University of Technology, Ridderprint Offsetdrukkerij B.V. Ridderkerk, the Netherlands, 219 p.

Snel, H., Schepers, J.G. (1992). Engineering models for dynamic inflow phenomena, *Wind Engineering and Industrial Aerodynamics*, vol. 39, pp. 267-281.

Snel, H., Schepers, J.G., editors (1995). *Joint Investigation of Dynamic Inflow Effects and Implementation of an Engineering Method*, Netherlands Energy Research Foundation ECN, Petten, the Netherlands, 326 p.

Song, Y.D., Dhinakaran, B., Bao, X.Y. (2000). Variable speed control of wind turbines using non-linear and adaptive algorithms, *Wind Engineering and Industrial Aerodynamics*, vol. 85, pp. 293-308.

Stapleton, S., Hopewell, P., Bryans, L. (2003). Dynamic models for modern wind turbine generators and their application to offshore wind farms, *In Third International Workshop on Large-Scale Integration of Wind Power and Transmission Networks for Offshore Wind farms*, Stockholm, Sweden, 12 p.

Svensson, J. (1998). *Grid-connected Voltage Source Converter – Control Principles and Wind Energy Applications*, Chalmers University of Technology, Göteborg, Sweden, Technical Report no. 331, 34 p.

Sobrink, K.H., Schettler, F., Bergmann, K., Stöber, R., Jenkins, N., Ekanayake, J., Pedersen, J.K., Pedersen, K.O.H. et all (1998). *Power Quality Improvements of Wind Farms*, Fredericia, Denmark, ISBN No.: 87-90707-05-2.

Sorensen, H.C., Hansen, J., Volund, P. (2001). Experience from the establishment of Middelgrunden 40 MW offshore wind farm, *In European Wind Energy Conference EWEC-2001*, Copenhagen, Denmark, pp. 541-544.

Sorensen, J.N., Kock, C.W. (1995). A model for unsteady rotor aerodynamics, *Wind Engineering and Industrial Aerodynamics*, vol. 58, pp. 259-275.

Sorensen, P., Hansen, A.D., Christensen, P., Mieritz, M., Bech, J., Bak-Jensen, B., Nielsen, H. (2003). *Simulation and Verification of Transient Events in Large Wind Power Installations*, Report Riso-R-1331(EN), Riso National Laboratory, Roskilde, Denmark, 80 p.

Taylor, C.W. (1994). *Power System Voltage Stability*, McGraw-Hill, Inc., New York, 273 p., in Chapter 3.

Vestas (2001), *OptiSpeed^{TM}, Vestas Converter System, General Edition*, Class 1, Item.no. 947543.R0, Vestas Wind Systems, Ringkobing, Denmark, 10 p.

Vestas. (2003). *Technical Description of the OptiSlip® Feature in Vestas Wind Turbines*, Class 1, Item no. 947525.R3, Vestas Wind Systems, Ringkobing, Denmark, 9 p.

Walker, J.F., Jenkins, N. (1997). *Wind Energy Technology*, John Wiley & Sons, London, U.K., 161 p.

Wu, X., Arumlampalam, A., Zha, C., Jenkins, N. (2003). Application of a static reactive power compensator (Statcom) and dynamic braking resistor (DBR) for the stability enhancement of a large wind farm, *Wind Engineering*, vol. 27, no. 2, pp. 93-106.

Xue, Y. (2005). *Development and implementation of an HVDC-VSC model for grid-connection of a large wind farm*, Electrical Engineering and Automation Group, Orsted DTU, Technical University of Denmark, 67 p.

Yamamoto, M., Motoyoshi, O. (1991) Active and reactive power control for doubly-fed wound rotor induction generator, *IEEE Transactions on Power Electronics*, vol. 6, no. 4, pp. 624-629.

Younsi, R., El-Batanony, I., Tritsch, J.B., Naji, H., Landjerit, B. (2001). Dynamic study of a wind turbine blade with horizontal axis, *European Journal of Mechanics A Solids*, vol. 20, pp. 241-252.

Zee, E. (2004). *Development of Offshore Wind Energy in Europe - Background document*, Policy Workshop organised by The Netherlands Ministry of Economic Affairs in co-operation with Concerned Action for Offshore Wind Energy Deployment, The Netherlands, Sept., 43 p.

Zhang, L., Watthanasarn, C., Shepherd, W. (1997). *Application of a matrix converter for the power control of a variable-speed wind-turbine driving a doubly-fed induction generator*, In IECON Proceedings, Industrial Electronics Conference, vol. 2, pp. 906-911.

Zinger, D.S., Muljadi, E. (1997). Annualized wind energy improvement using variable speeds, *In IEEE Annual Meeting: Industrial and Commercial Power Systems Technical Conference*, Conference Records, pp. 80-83.

Oye, S. (1986). Unsteady wake effects caused by pitch –angle changes, *In IEA R&D WECS Joint Action on Aerodynamics of Wind Turbines, 1st Symposium*, London, U.K., pp.58-79.